Composite and Nanocomposite Materials - From Knowledge to Industrial Applications

Edited by Tri-Dung Ngo

Published in London, United Kingdom

IntechOpen

Supporting open minds since 2005

Composite and Nanocomposite Materials - From Knowledge to Industrial Applications
http://dx.doi.org/10.5772/intechopen.80186
Edited by Tri-Dung Ngo

Contributors
Beera Satish Ben, B Avinash Ben, Abdelhak Khechai, Mohamed-Ouejdi Belarbi, Abdelouahab Tati, P.M. Mohite, Jinhua Wei, Bin Lin, Haoji Wang, Tri-Dung Ngo, Venkatachalam N, Mikko Kanerva, Christian Emeka Okafor, Ihueze Christopher Chukwutoo, Malek Adnan Alghdeir, Khaled Mayya, Mohamed Dib, S Kalaiselvan, J Manivannan, R Padmavathi, Ick Soo Kim, Phan Duy Nam, Xian Zhang, Yan Xu, Rajendra Sukhadeorao Dongre

First published in London, United Kingdom, 2020 by IntechOpen
IntechOpen is the global imprint of INTECHOPEN LIMITED, registered in England and Wales, registration number: 11086078, 7th floor, 10 Lower Thames Street, London, EC3R 6AF, United Kingdom
Printed in Croatia

British Library Cataloguing-in-Publication Data
A catalogue record for this book is available from the British Library

Additional hard and PDF copies can be obtained from orders@intechopen.com

Composite and Nanocomposite Materials - From Knowledge to Industrial Applications
Edited by Tri-Dung Ngo
p. cm.
Print ISBN 978-1-78985-390-2
Online ISBN 978-1-78985-461-9
eBook (PDF) ISBN 978-1-78985-462-6

For EU product safety concerns:
IN TECH d.o.o., Prolaz Marije Krucifikse Kozulić 3, 51000 Rijeka, Croatia, info@intechopen.com or visit our website at intechopen.com.

Meet the editor

Dr. Ngo graduated with an undergraduate degree in Chemical Engineering fron Polytechnic University (HoChiMinh University of Technology), Vietnam, in 1997. He received a master's degree in Chemical Engineering in 2001. In 1997, he held a position as lecturer and researcher in the Department of Chemical and Engineering at Polytechnic University. In 2001, he joined the Department of Materials Engineering at Polytechnic University as lecturer and researcher. He obtained his PhD in Mechanical Engineering from Concordia University, Montreal, Canada, in 2007. From 2003 to 2007, he worked as invited researcher at Industrial Materials Institute (IMI) of National Research Council of Canada (NRC). After completing his Natural Sciences and Engineering Research Council of Canada (NSERC) post-doctoral fellowship at NRC in 2008, he joined IMI of NRC as researcher. Dr. Ngo is currently working at InnoTech Alberta (formerly Alberta Research Council (1921-2010) and Alberta Innovates-Technology Futures (2010-2016)), a subsidiary of Alberta Innovates, as senior research scientist for Bio-Industrial Services, a position he has been appointed in late 2014. His research focuses on biomass processing and development of novel, eco-friendly applications using renewable and sustainable biomaterials including but not limited to polymers, biopolymers, composites, biocomposites, nanocomposites, films, additive manufacturing, celluloses, fibers, nanocelluloses, and lignin for industrial applications. Dr. Ngo is also an adjunct professor in the Department of Civil and Environmental Engineering at University of Alberta, Edmonton, Canada. Dr. Ngo has been acting as a project manager for numerous industrial and academic projects at both the national and international levels. Dr. Ngo has authored about 100 papers published in international conference proceedings and international journals. He is an author of two books and nine scientific book chapters. He has written many proposals, business plans, and scientific reports. He is a reviewer for several international journals and scientific associations.

Contents

Preface

Composite materials have played an important role throughout human history, from housing early civilizations to enabling future innovations. Composites offer many benefits, key among them being corrosion resistance, design flexibility, durability, light weight, and strength. Composites have permeated our everyday lives through products that are used in construction, medical applications, oil and gas, transportation, sports, aerospace, consumer products, and much more. Over the last decades, nanotechnology has become a rapidly evolving technology. As such, science, engineering, and technology have merged to bring nanoscale materials that much closer to reality. This is one of the fastest growing areas for research. Nanocomposite materials are helping improve products that we use every day and creating new, exciting products for the future. Composites and nanocomposites composed of reinforcements, nano-reinforcements, and matrices are well-known engineering materials. Various reinforcements and nano-reinforcements used in the materials have different properties and thus affect their properties in different ways.

Keeping in mind the advantages of composite and nanocomposite materials, this book covers fundamental affects, product developments, and applications of the materials including material chemistry, designing, manufacturing, properties, and utilization of the materials in various applications. The book consists of 12 chapters that summarize recently developed research concerning composites and nanocomposites. Each chapter gives an overview of a particular material, its processing, and successful utilization for selected applications. The book also summarizes the recent developments made in the area of advanced composite and nanocomposite materials. Several critical issues and suggestions for future work are discussed, underscoring the roles of researchers for the efficient development of composites and nanocomposites through value additions to enhance their use.

Chapter 1, "Introduction to composite materials," addresses the advantages of composite materials as well as fundamental effects, product developments, and applications of composites, including material chemistry, designing, manufacturing, properties, and utilization of the materials in various applications. Composites offer many benefits and have permeated our everyday lives; however, to facilitate the advantages of the composites, several aspects must be considered: (a) concept development, (b) material selection and formulation, (c) material design, (d) product manufacturing, (e) market, and (f) regulations.

Chapter 2, "Introduction of natural fiber composite using in paperboard industry," addresses the mechanical properties, chemical compositions, crystallinity, and thermal stability of the newly developed composites. The results are analyzed to ensure the suitability of natural fiber as an optional fiber-material for the paperboard industry.

Chapter 3, "Advanced treatments of aramid fibers for composite laminates," is an overview of the aramid fibers, an important group of fibers for composite applications. This chapter discusses recent studies in micro-testing, for example, in the form of micro-droplet tests. Surface treatments and fiber modifications of the aramid fibers are also discussed. Furthermore, the properties of the resin, fiber, and interface are presented.

Chapter 4, "Vapor-Grown Carbon Fiber Synthesis, Properties, and Applications" discusses synthesis and properties of carbon fibers and vapor-grown carbon nanofibers as well as their mechanical, electrical, and thermal properties. In addition, several potential applications of fibers, such as in composites and batteries, are also discussed. Carbon nanofibers and their composites can be utilized in numerous fields, including sensors, electrode materials, and electromagnetic shielding. Electrode materials and the realization of extraordinary structural designs to ensure large specific areas without sacrificing mechanical performances are key factors in enhancing the performances of these materials.

Chapter 5, "Fiber composites made of low-dimensional carbon materials," provides an overview of the research on carbon nanotubes (CNTs), carbon fibers (CFs), and graphene-based fibers (GBFs) in composite materials. It also summarizes the main preparation methods, properties, and application fields of CNTs, CFs, and GBFs. Through various methods of modification, researchers continue to prepare CNT composite nanomaterials with excellent performance. Finally, the chapter examines existing problems and future development trends of carbon-based composites.

Chapter 6, "Reinforce Fabricated Nano-Composite Matrixes for Modernization of S & T in New Millennium" is a brief review of certain recent advances in assembly and reconfiguration of polymeric nanostructure composites and highlights their role in computer simulation. In addition, it presents the fundamental principles of assembly science as well as provides critical design tools for assembly engineering of complex nanostructured materials. The chapter also discusses the advantages of nanocomposites.

Chapter 7, "Composite nanofibers: recent progress in adsorptive removal and photocatalytic degradation of dyes," reviews the state of the art of a new generation of nanomaterials based on electrospun composite nanofibers for dye removal from wastewater. Natural polymers-based nanofibers, nanofibers with unique morphology, and carbon nanofibers are reviewed. The various nanostructures, adsorption capacity, advantages, and drawbacks are discussed along with mechanistic actions in the adsorption process and photocatalytic performance. The chapter covers multiple intriguing topics with in-depth discussion. This is a valuable reference for researchers who are working on nanomaterials and the treatment of colored waters.

Chapter 8, "Nanosilica composite for greenhouse application," presents the results of experiments on silicon dioxide insulation materials mixed with low-density polyethylene (LDPE) at a different proportion to prevent the transmittance of IR domain and to allow the transmittance of UV–Vis domains. Several ratios of nanosilica particles are employed to fabricate LDPE composites using melt mixing and hot molding methods. The mechanical properties of nanocomposites such as tensile strength are evaluated and discussed. The advantages of using nanosilica in the nanocomposite materials are summarized and the use of these LDPE/silica nanocomposites to build a mini greenhouse is presented.

Chapter 9, "Damage identification and assessment using lamb wave propagation parameters and material damping in reinforced polymer (FRP) composite laminates," presents a methodology to identify damage in fiber-reinforced polymer (FRP) composites. The Lamb wave dispersion theory and loss less finite element model to find the fitted peak frequency and the modal frequencies in the composite laminates, respectively, are provided. The proposed method is implemented on FRP laminates (CFRP and GFRP) and the results are compared with bandwidth method.

Chapter 10, "Strength improvement and stress analysis of E-glass laminated plates with circular notches using digital image correlation," addresses the stress concentration and tensile strength degradation of E-glass/epoxy laminates through a combination of both experimental and numerical studies. The numerical study using the finite element method (FEM) is discussed. The main aim of this work is to improve the ultimate strength of perforated composite plates, by using the defense hole system (DHS) technique. A digital image correlation (DIC) technique is also used to get the full-field surface strain measurements in perforated samples with various open hole diameters and DHS configurations, in order to show their effects on failure strength.

Chapter 11, "Strength analysis and variation of elastic properties in plantain fiber/polyester composites for structural applications," assesses the extent of variation of elastic properties (E_x, E_y, G_{xy}, v_{xy}, v_{yx}, m_x, m_y) with fiber orientation using MATLAB functions while considering the extent of variation of the tangential stresses around an idealized functional hole edge. The utilization of plantain fiber-reinforced composites in structural applications empowers architects to make great accomplishments in the usefulness, security, and economy of development. The chapter also presents some notable design procedures in handling such limitations in plantain fiber-reinforced composites using relevant failure theories.

Chapter 12, "Surface measurement and evaluation of fiber woven composites," presents a grading surface measurement and evaluation system for WCMCs based on their microstructures. The system includes four levels: fiber, fiber bundle, cell body, and the whole surface. On the fiber level, the typical forms of fiber damage and their effects on the surface morphology of WCMCs are analyzed, which lays a foundation for the measurement and evaluation methods on the next three levels. On each subsequent level, the system proposes a set of surface measurement-sampling parameter determination methods and surface quality evaluation methods based on the principle of statistics.

As the editor, I have attempted to compile, unify, and present the emerging research trends in composite and nanocomposite materials. I hope this book will contribute to the advancement of both science and technology in these exciting areas. It is highly recommended for graduate students, engineers, researchers/faculty, and professionals/practitioners in different fields such as chemistry, chemical engineering, civil engineering, forestry and agriculture, pulp and paper producers, and materials sciences.

Acknowledgments

I would like to thank all contributors, who provided excellent contributions and commitments on the chapters and discussions for this book. I wish to express my appreciation to my family, my colleagues, and friends for the support they provided me. I would like to extend my appreciation to the publishing staff at IntechOpen, Ms. Dajana Pemac and Ms. Anja Filipovic, for their valuable support during the publishing of this book.

Tri-Dung (T.-D.) Ngo, Ph.D.
Senior Research Scientist,
Adjunct Professor,
Bio-Industrial Services,
InnoTech Alberta,
Edmonton, Alberta, Canada

Department of Civil and Environmental Engineering,
Faculty of Engineering,
University of Alberta,
Edmonton, Alberta, Canada

Section 1

Composite Materials

Chapter 1

Introduction to Composite Materials

Tri-Dung Ngo

Abstract

Composite materials have played an important role throughout human history, from housing early civilizations to enabling future innovations. Composites offer many benefits; the key among them are corrosion resistance, design flexibility, durability, light weight, and strength. Composites have permeated our everyday lives such as products that are used in constructions, medical applications, oil and gas, transportation, sports, aerospace, and many more. Some applications, such as rocket ships, probably would not get off the ground without composite materials. This chapter addresses the advantages of fibre composite materials as well as fundamental effects, product development, and applications of fibre composites, including material chemistry, designing, manufacturing, properties, and utilisation of the materials in various applications.

Keywords: fibre, matrix, composite, fibre composite, thermoset, thermoplastic, natural fibre, biopolymer

1. Introduction

Composites exist in nature. A piece of wood is a composite, with long cellulose fibres held together by a substance called lignin. Composite materials are formed by combining two or more materials that have quite different properties, and they do not dissolve or blend into each other. The different materials in the composite work together to give the composite unique properties. Humans have been using composite materials for thousands of years in different areas. The first uses of composites date back to the 1500 BC, when early Egyptians and Mesopotamian settlers used a mixture of mud and straw to create strong and durable buildings. The combination of mud and straw in a block of brick provides it a strong property against both squeezing and tearing or bending. The straw continued to provide reinforcement to ancient composite products, including pottery and boats [1]. In 1200 AD, the Mongols invented the first composite bow using a combination of "animal glue", bone, and wood. The bows were pressed and wrapped with birch bark. These bows were powerful and accurate. Composite Mongolian bows helped to ensure Genghis Khan's military dominance. Due to their advantages such as being light weight and strong, many of the greatest advancements in composites were the result of wartime needs. During World War II, many composite materials were developed and moved from the laboratory into actual production [1, 2].

The development and need for composite materials also result in the fibre-reinforced polymers (FRP) industry. By 1945, more than 7 million pounds of glass fibres were used for various products, primarily for military applications. Composite materials continued to take off after the war and grew rapidly through the 1950s. The composite innovators were ambitiously trying to introduce composites into other markets such as aerospace, construction, and transportation. Soon the benefits of FRP composites, especially its corrosion resistance, became known to the public sector. Boats were one obvious product that benefited. The first composite commercial boat hull was introduced in 1946. A full automobile body was made from composite and tested in 1947 [1, 2]. This led to the development of the 1953 Chevrolet Corvette. The advent of the automobile age gave rise to several new methods for moulding such as compression moulding of bulk moulding compound (BMC) and sheet moulding compound (SMC). The two techniques emerged as the dominant method of moulding for the automotive industry and other industries. In the early 1950s, manufacturing methods such as large-scale filament winding, pultrusion, and vacuum bag moulding were developed. In the 1960s, the marine market became the largest consumer of composite materials [1, 2]. In 1961, the first carbon fibre was patented and several years later became commercially available. In the 1970s the composites industry began to mature. Many better resins and improved reinforcing fibres were developed during this period for composite applications. In the 1970s, the automotive market surpassed marine as the number one market—a position it retains today. During the late 1970s and early 1980s, composites were first used in infrastructure applications in Asia and Europe. The first all-composites pedestrian bridge was installed in Aberfeldy, Scotland, in the 1990s. In this period, the first FRP-reinforced concrete bridge deck was built in McKinleyville, West Virginia, and the first all-composites vehicular bridge deck was built in Russell, Kansas. Composites continue to find applications today [1–3]. Nanomaterials are incorporated into improved fibres and resins used in new composites. Nanotechnology began to be used in commercial products in the early 2000s. Bulk carbon nanotubes can be used as composite reinforcement in polymers to improve the mechanical, thermal, and electrical properties of the bulk product [3].

Nowadays, the composite industry is still evolving, with much of the growth now focused around renewable energy. Wind turbine blades, especially, are constantly pushing the limits on size and require advanced composite materials, for example, the engineers can design to tailor the composite based on the performant requirements, making the composite sheet very strong in one direction by aligning the fibres that way, but weaker in another direction where strength is not so important. The engineers can also select properties such as resistance to heat, chemicals, and weathering by choosing an appropriate matrix material. In recent years, an increasing environmental consciousness and awareness of the need for sustainable development have raised interest in using natural fibres as reinforcements in composites to replace synthetic fibres [4–7]. This chapter seeks to provide an overview of the science and technology in relation to the composite material, manufacturing process, and utilisation.

2. Polymer matrix

In general, a composite consists of three components: (i) the matrix as the continuous phase; (ii) the reinforcements as the discontinuous or dispersed phase, including fibre and particles; and (iii) the fine interphase region, also known as the interface [8, 9]. By carefully choosing the matrix, the reinforcement, and the manufacturing process that brings them together, the engineers can tailor the properties to meet specific requirements [10]. Over the recent decades, many new composites have been developed, some with very valuable properties.

Any material can serve as a matrix material for composite. However, matrix materials are generally ceramics, metals, and polymers. In reality, the majority of matrix materials that exist on the composites market are polymer. There are several different polymer matrices which can be utilised in composite materials. Among the polymer matrix composites, thermoset matrix composites are more predominant than thermoplastic composites. Though thermoset and thermoplastics sound similar, they have very different properties and applications. Understanding the performance differences can help to make better sourcing decisions and the product designs as composites [11].

Thermosets are materials that undergo a chemical reaction or curing and normally transform from a liquid to a solid. In its uncured form, the material has small, unlinked molecules known as monomers. The addition of a second material as a cross-linker, curing agent, catalyst, and/or the presence of heat or some other activating influences will initiate the chemical reaction or curing reaction. During this reaction, the molecules cross-link and form significantly longer molecular chains and cross-link network, causing the material to solidify. The change of the thermoset state is permanent and irreversible. Subsequently, exposure to high heat after solidifying will cause the material to degrade, not melt. This is because these materials typically degrade at a temperature below where it would be able to melt.

Thermoplastics are melt-process able plastics. The thermoplastic materials are processed with heat. When enough heat is added to bring the temperature of the plastic above its melting point, the plastic melts, liquefies, or softens enough to be processed. When the heat source is removed and the temperature of the plastic drops below its melting point, the plastic solidifies back into a glasslike solid. This process can be repeated, with the plastic melting and solidifying as the temperature climbs above and drops below the melting temperature, respectively. However, the material can be increasingly subject to deterioration in its molten state, so there is a practical limit to the number of times that this reprocessing can take place before the material properties begin to suffer. Many thermoplastic polymers are addition-type, capable of yielding very long molecular chain lengths or very high molecular weights [12].

Both thermoset and thermoplastic materials have its place in the market. In broad generalities, thermosets tend to have been around for a long time and have a well-established place in the market, frequently have lower raw material costs, and often provide easy wetting of reinforcing fibre and easy forming to final part geometries. In other words, thermosets are often easier to process than thermoplastic. Thermoplastics tend to be tougher or less brittle than thermoset. They can have better chemical resistance, do not need refrigeration as uncured thermosets (prepreg materials) frequently do, and can be more easily recycled and repaired. **Table 1** presents a comparison between thermoset and thermoplastic. This table is not providing all but rather some information for the researchers and manufacturers when considering the utilisation of these materials.

Thermosets are classified into polyester resins, epoxy resins, vinyl ester resins, phenolic, polyurethane, and other high-temperature resins such as cyanate esters, etc. The rapid industrialisation in developing economies the world over is one of the major boosting factors for the thermoset market. The demand for high-performance and lightweight materials from various end-use industries such as automotive, chemical tanks, and water tanks is expected to expand the global market for thermosets over the next 6 years. The growing demand for thermosets from emerging economies like Brazil, Russia, India, and China (BRIC) is expected to drive the market. BRIC nations are the four fastest-growing economies in the world with their GDP growth rates higher than the global GDP growth rate.

	Thermoset	Thermoplastic
Processing	Contain monomers that cross-link together during the curing process to form an irreversible chemical bond. The cross-linking process eliminates the risk of the product remelting when heat is applied, making thermosets ideal for high-heat applications such as electronics and appliances	Pellets soften when heated and become more fluid as additional heat is applied. This characteristic allows thermoplastics to be remoulded and recycled without negatively affecting the material's physical properties
Features and benefits	• There are multiple thermoset resins that offer various performance benefits • Significantly improve the material's mechanical properties, providing enhances chemical resistance, heat resistance, and structural integrity. Thermoset are often used for sealed products due to their resistance to deformation • Cannot be recycled • Cannot be remoulded or reshaped (not melt if heat) • Easy to wet the reinforcing fibres and fillers • More resistant to high temperatures than thermoplastics • Highly flexible design • Thick to thin wall capabilities • Excellent aesthetic appearance • High levels of dimensional stability • More difficult to surface finish • Cost-effective	• There is multiple thermoplastic that offer various performance benefits • Commonly offer high strength, shrink-resistance, and easy bendability. Depending on the polymers, thermoplastics can serve low-stress applications such as plastic bags or high-stress mechanical parts • Highly recyclable • Can melt if heated, remoulding/reshaping capabilities • More difficult to wet the reinforcing fibres and fillers • High-impact resistance • Chemical resistant • Hard crystalline or rubbery surface options • Aesthetically superior finishes • Ecofriendly manufacturing • Generally, more expensive than thermoset

Table 1.
Thermoset vs. thermoplastic.

However, frequent fluctuation in raw material prices acts as one of the major factors inhibiting the market growth. Asia-Pacific accounts for the biggest market for thermosets owing to the growth of the automobile market, primarily in China and India. Japan is a mature market and is expected to remain stagnant over the next years. China is the biggest automobile market in the world, and India also lists itself in the top five automobile markets in the world. Asia, along with being the largest market, is also the fastest-growing market for thermosets. The North American market for thermosets is primarily driven by the regulatory initiative to reduce automobile weight by 50% by 2020 in the USA in order to cut fuel consumption. Polyester resins and polyurethane account for the two most popular types of thermosets in the global market. The global market for thermosets is dominated by big multinational corporations which are present across the value chain. Some of the major companies operating in the thermosets market include Arkema, BASF, Asahi Kasei Chemical Corp, Bayer AG, Chevron Phillips Chemical Company LLC, Sinopec, Dow Chemical Company, Eastman Chemical Company, and Lyondell Basell Industries, among others [13]. To date, thermosets have been used predominantly in the industry. Thermosets are generally favoured for a variety of reasons, especially on commercial aircraft. Thermoset composites have been used for 30–40 years in aerospace. For example, the fuselage of the Boeing 787 is an epoxy-based polymer [14].

On the other hand, the use of thermoplastic polymers (acrylic, polyolefin, acrylonitrile butadiene styrene (ABS), etc.), the more easily moldable and resettable composite material relative to thermoset polymers, is a growing material trend in the fibre-reinforced polymer (FRP) industry. According to the American Composites Manufacturers Association (ACMA), the thermoplastic industry is expected to grow 4.9% over the next years and reach an estimated $8.2 billion by 2017, with even larger opportunities in emerging economies. Thermoplastic polymers also offer an easy solution to recycling composite components, a concern when it comes to adopting composite materials. Thermoplastic composites can repeat the heating and cooling cycle many times, thus giving the product an almost indefinite shelf life and adding more value for industries concerned with composite recyclability. This is especially the case for the growth of natural fibre thermoplastics in the USA and Western Europe. For example, wood-plastic composites, used for decking material and other wood substitutions, have grown by 35–40% in the past 5 years. According to Lucintel (the premier global management consulting and market research firm), countries in Asia and Eastern Europe will lead the growth for thermoplastic adoption because automotive production and thermoplastic automotive component production are quickly growing in those regions. However, the automotive sectors in the USA and Western Europe may not experience the same high rate of growth but are expected to develop steadily in the next 5 years, mainly due to the acceptance of new composite application. The study indicates that although gains will be limited by rising energy costs and competition from lower cost materials, there is significant opportunity in emerging economies such as China, Russia, Brazil, and India [15]. Recently, a major trend in the aerospace industry is a move toward greater use of thermoplastics vs. "traditional" thermoset epoxies. This also opens an opportunity for thermoplastics.

Thermoplastic are the dominant plastic materials overall, especially in non-reinforced applications. Thermosets are used in non-reinforced applications for a specific purpose where they have an advantage because of some unique property. However, within the reinforced or composites marketplace, thermoset dominant and thermoplastic are used only in applications where their unique advantages are important. Within the composite market, thermoset represents about 80% of the total material used [16]. The global composite resin market size by end-use applications, in terms of value, was USD 9317.4 Million in 2014 and is projected to grow at a CAGR of 5.6% between 2015 and 2020 [17].

As mentioned above, thermoplastics are capable of being repeatedly softened by the application of heat and hardened by cooling and have the potential to be the most easily recycled, which has seen them most favoured in recent commercial uptake, whereas better realisation of the fibre properties is generally achieved using thermosets. There are several types of polymers in the market. The most common polymers are summarized in **Table 2** [18–23].

3. Reinforcements

Composite reinforcements can be in various forms such as fibres, flakes, or particles. Each of these has its own properties which can be contributed to the composites, and therefore, each has its own area of applications. Among the forms, fibres are the most commonly used in composite applications, and they have the most influence on the properties of the composite materials. These reasons are that the fibres have the high aspect ratio between length and diameter, which can provide effective shear stress transfer between the matrix and the fibres, and the ability to process and manufacture the composites part in various shapes using different techniques.

	Polymers	Density (g/cm^3)	Elongation (%)	Tensile strength (MPa)	Young's modulus (GPa)
Thermoplastic	Acrylonitrile styrene acrylate (ASA)	1.0–1.1	30.0	43.5	2.2
	Acrylonitrile butadiene styrene (ABS)	1.0–1.1	270.0	47.0	2.1
	Cross-linked polyethylene (PE)	0.9	350.0	18.0	0.5
	Ethylene vinyl acetate (EVA)	0.9–1.0	750.0	17.0	0.02
	High-density polyethylene (HDPE)	0.9–1.0	150.0	32.0–38.2	1.3
	High-impact polystyrene (HIPS)	1.0	2.5	42.0	2.1
	Low-density polyethylene (LDPE)	0.9	400.0	10.0–11.6	0.2–0.3
	Nylon 6 (PA 6)	1.1	60.0	81.4	2.8
	Nylon 66 (PA 66)	1.1	60.0	82.7	2.8
	Perfluoroalkoxy (vinyl ether)	2.15	260.0–300.0	28.0–31.0	0.50–0.60
	Polybutylene (PB)	0.95	220–300	29.0–35.0	0.29–0.30
	Polylactic acid (PLA)	1.2–1.3	2.1–30.7	5.9–72.0	1.1–3.6
	Polycarbonate (PC)	1.2	200.0	69.0	2.3
	Polycaprolactone (PCL)	1.1	700.0	16.0–23.0	0.4
	Polyethylene cross-linked (PEX)	0.92		20.0	
	Polyethylene terephthalate (PET)	1.5–1.6	300.0	55.0–159.0	2.3–9.0
	Polyether ether ketone (PEEK)	1.3–1.5	1.6–50.0	92.0–95.0	3.7–24.0
	Polyether ketone (PEK)	1.2–1.4	20.0	100.0–110.0	3.5
	Polyhydroxyalkanoates (PHA)	1.2–1.3	2.0–1200.0	10.0–39.0	0.3–3.8
	Polyhydroxybutyrate (PHB)	1.2	1.56–6.0	24.0–40.0	3.5–7.7
	Poly-3-hydroxybutyrate (P-3-HB)	1.3	0.4–6.0	40.0	3.5
	Poly-3-hydroxybutyrate-co-3-hydroxyvalerate (P-3-HB-3 HV)	0.2–0.3	1.6–20.0	23.0–40.0	3.5
	Poly-3-hydroxybutyrate (P-3-HB)	1.2	1000.0	104.0	—
	Poly(methyl methacrylate) (PMMA)	1.1–1.2	2.5	72.4	3.0
	Polypropylene (PP)	0.9–1.3	80.0	35.8	1.6
	Polystyrene (PS)	1.04	1.6	34.0	3.0
	Polytetrafluoroethylene (PTFE)	2.20	40.0–650.0	0.862–41.4	0.392–2.25
	Polyvinyl chloride (PVC)	1.3–1.5	50.0–80.0	52.0–90.0	3.0–4.0
	Polyvinylidene fluoride (PVDF)	1.8	50.0	43.0	2.0
	Rigid thermoplastic Polyurethane (RTPU, PUR-RT)	1.1	5.0	75.0	4.0

	Polymers	Density (g/cm^3)	Elongation (%)	Tensile strength (MPa)	Young's modulus (GPa)
Thermoset	Epoxy (EP)	1.2–1.3	1.3	55.0–130.0	2.7–4.1
	Melamine formaldehyde (MF)	1.5–1.6	0.6	65.0	12.0
	Phenol formaldehyde (PF)	1.2	1.2	45.0–60.0	4.0–7.0
	Rigid thermoset polyurethane (RPU)	1.2	90.0	60.0	2.2
	Unsaturated polyester (UPE)	1.1	2.0	34.0–105.0	2.1–3.5
	Urea formaldehyde (UF)	1.5–1.6	0.8	65.0	9.0
	Polyurethane rubber	1.2–1.3	300.0–580.0	39.0	2.0–10.0
	Vinyl ester (VE)	1.23	2.0–12.0	73.0–81.0	3.0–3.5

Table 2.
Properties of some polymers.

Various types of fibres have been utilised to reinforce polymer matrix composites. The most common are carbon fibres (AS4, IM7, etc.), glass fibre (E-glass, S-glass, etc.), aramid fibres (Kevlar® and Twaron®), and boron fibres. Glass fibres have been used as reinforcement for centuries, notably by Renaissance Venetian glass workers. Commercially important continuous-glass fibre filaments were manufactured in 1937 by a joint venture between Owens-Illinois and Corning Glass. A variety of glass fibre compositions are available for different purposes as presented below. **Table 3** shows compositions of some commonly used glass fibres for composite materials.

- Grade A is high alkali grade glass, originally made from window glass.
- Grade C is chemical-resistant grade glass for acid environments or corrosion.
- Grade D is low dielectric grade glass, good transparency to radar (quartz glass).
- Grade E is electrical insulation grade; this is the most common reinforcement grade.
- Grade M is high modulus grade glass.
- Grade R is reinforcement grade glass; this is the European equivalent of S-glass.
- Grade S is high strength grade glass, a common variant is S2-glass. This fibre has higher Young's modulus and temperature resistance than E-glass. It is also significantly more expensive.

Table 4 presents the mechanical properties of the main grades of glass fibre for composite materials.

Carbon fibre was first invented near Cleveland, Ohio, in 1958. It wasn't until a new manufacturing process was developed at a British research centre in 1963 that carbon fibre's strength potential was realised [27]. The principle precursors for carbon fibres are polyacrylonitrile (PAN), pitch, cellulose (Rayon), and some other potential precursors such as lignin and polyethylene. Carbon fibres are

Oxide	E-glass with boron	E-glass without boron	ECR-glass	S-2 glass	R-glass	Effect on fibre properties
SiO_2	52–56	59	54–62	64–66	60–65	Very low thermal expansion
Al_2O_3	12–16	12.1–13.2	9–15	24–26	17–24	Improved chemical durability
B_2O_3	5–10	—	—	—	—	Low thermal expansion
CaO	16–25	22–23	17–25	—	5–11	Resistance to water, acids, and alkalis
MgO	0–5	3.1–3.4	0–5	8–12	6–12	Resistance to water, acids, and alkalis
ZnO	—	—	2.9	—	—	Chemical durability
Na_2O	0–1	0.6–0.9	1.0	0–0.1	0–2	High thermal expansion, moisture sensitivity
K_2O	Trace	0–0.2	0.2	—	0–2	High thermal expansion, moisture sensitivity
TiO_2	0.2–0.5	0.5–1.5	2.5	—	—	Improved chemical durability especially alkali resistance
Zr_2O_3	—	—	—	0–1	—	—
Li_2O	—	—	—	—	—	High thermal expansion, moisture sensitivity
Fe_2O_3	0.2–0.4	0.2	0.1	0–0.1	—	Green colouration
F_2	0.2–0.7	0–0.1	Trace	—	—	—

Table 3.
Composition for some commonly used glass fibres [24–26].

Fibre	Density (kg/m^3)	Young's modulus (GPa)	Virgin filament strength (MPa)	Roving strength (MPa)	Strain to failure (%)
A (alkali)	2460	73	3100	2760	3.6
C (chemical)	2460	74	3100	2350	~
D (dielectric)	2140	55	2500	~	~
E (electrical)	2550	71	3400	2400	3.37
R (reinforcement)	2550	86	4400	3100	5.2
S (strength)	2500	85	4580	3910	4.6
S_2	2460	90	3623	~	~
S_3	2830	99	3283	~	~

Table 4.
Mechanical properties of the main grades of glass fibre [24].

manufactured by stretching PAN polymer precursor, melt spinning of molten pitch, and graphitization under tensile stress [28].

The modulus of carbon fibres depends on the degree of perfection of the alignment. Imperfections in alignment results in complex shaped voids elongated parallel to the fibre axis, which act as stress raisers and points of weakness. The alignment varies considerably with the manufacturing route and conditions. High-modulus fibres are those which have been subjected to heat treatment in excess of 1650°C, possess three-dimensional ordering of the atoms, have carbon contents above 99% (although their graphitic structure is still less than 75%), and have a tensile modulus above 350 GPa. High-modulus, high-strength carbon fibres have diameters of 7–8 μm and consist of small crystallites of "turbostratic" graphite. The layers have no regular stacking sequence, and the average spacing between the planes is 0.34 nm. To obtain high modulus and strength, the layer planes of the graphite must be aligned parallel to the fibre axis [29]. Carbon fibres have several advantages including high stiffness, high tensile strength, low weight, high chemical resistance, and high temperature. The carbon fibres can be utilised in various applications such as aerospace, automotive, sporting goods, and consumer goods. **Table 5** shows properties for the different grades of carbon fibre.

Kwolek is a DuPont chemist who in 1965 invented an aramid fibre known as Kevlar, the lightweight, stronger-than-steel fibre used in bulletproof vests and other body armour around the world. The chemical structure of the materials is being alternated aromatic (aryl) benzene rings and the amide (CONH) group. The commercial name of the reinforcement's fibres is Kevlar from DuPont and Twaron from AkzoNobel, which are believed to be poly-(para-phenylene terephthalamide). The polymer is produced by the elimination of hydrogen chloride from terephthaloyl chloride and para-phenylene diamine. The polymer is washed and dissolved in sulphuric acid to form a partially oriented liquid crystal solution. The solution is spun through small die holes, orientation taking place in the spinnerette, and the solvent is evaporated. Hull suggests that the solution is maintained between −80°C and −50°C before spinning and is extruded into a hot-walled cylinder at 200°C. Kevlar was introduced for commercial products in 1971. There are three principal types of Kevlar fibre as shown in **Table 6**.

Recently, with advantages of reasonable mechanical properties, low density, environmental benefits, renewability, and economic feasibility, natural fibres have been paid more attention to in composite applications. The natural fibres in simple definition are fibres that are not synthetic or man-made and are categorized based

Precursor	PAN	PAN	Pitch	Pitch	Rayon	Pitch (K13D2U)
Modulus	Low	High	Low	High	Low	Ultrahigh
Tensile modulus (GPa)	231	392	161	385	41	931
Tensile strength (GPa)	3.4	2.5	1.4	1.8	1.1	3.7
Strain to failure (%)	1.4	0.6	0.9	0.4	2.5	0.4
Relative density	1.8	1.9	1.9	2.0	1.6	2.2
Carbon assay (%)	94	100	97	99	99	>99

Table 5.
Indicative properties for the different grades of carbon fibre [27].

Fibre type		E (GPa)	σ' (GPa)	ε' (%)
Kevlar 29	High-toughness, high-strength, intermediate modulus for tyre cord reinforcements	83	3.6	4.0
Kevlar 49	High modulus high-strength for composite reinforcement	131	3.6	2.8
Kevlar 149	Ultra-high modulus recently introduced	186	3.4	2.0

Table 6.
Characteristics of the different grades of aramid fibre [27].

on their origin from animals, mineral, or plant sources [30]. Natural fibres are one such proficient material which would be utilised to replace the synthetic materials and their related products for the applications requiring less weight and energy conservation. Natural plant fibres are entirely derived from vegetative sources and are fully biodegradable in nature. Fibre-reinforced polymer matrix got considerable attention in numerous applications because of its good properties. The current indicators are that interest in natural fibre composites by the industry will keep growing quickly around the world. The application of natural fibre-reinforced polymer composites and natural-based resins for replacing existing synthetic polymer or glass fibre-reinforced materials is huge. However, natural fibre quality is influenced significantly by the age of the plant, species, growing environment, harvesting, humidity, quality of soil, temperature, and processing steps, and there is a move to reduce the on-field processing to improve consistency and reduce costs. The properties of several natural fibres and commonly used synthetic fibres are shown in **Table 7** [31–35].

Fibre	Density (g/cm^3)	Elongation (%)	Tensile strength (MPa)	Young's modulus (GPa)
Abaca	1.5	—	511.0–1051.0	13.5–29.8
Alfa	0.89	—	350.0	22.0
Bagasse	1.2	1.1	20.0–290.0	19.7–27.1
Banana	1.3–1.4	2.0–7.0	54.0–789.0	3.4–32.0
Bamboo	1.5	—	575.0	27.0
Coconut	1.4–3.8	—	120.0–200.0	19.0–26.0
Coir	1.2	15.0–30.0	175.0–220.0	4.0–6.0
Cotton	1.5–1.6	3.0–10.0	287.0–597.0	5.5–12.6
Curaua	1.4	—	825.0	9.0
Flax	1.4–1.5	1.2–3.2	345.0–1500.0	27.6–80.0
Hemp	1.4–1.5	1.6	550.0–900.0	70.0
Henequen	1.4	3.0–4.7	430.0–580.0	—
Isora	1.2	—	550.0	—
Jute	1.3–1.5	1.5–1.8	393.0–800.0	10.0–30.0
Kapok	0.4	—	93.3	41.0
Kenaf	1.2	2.7–6.9	295.0	—
Palf	1.4	3.0	170.0–635.0	6.2–24.6
Piassava	1.4	—	138.5	2.8
Pineapple	1.5	1.0–3.0	170.0–1672.0	82.0
Ramie	1.5	2.0–3.8	220.0–938.0	44.0–128.0
Silk	1.3–1.4	—	650.0–750.0	16.0
Sisal	1.3–1.5	2.0–14.0	400.0–700.0	9.0–38.0
Softwood Kraft	1.5	—	1000.0	40.0
Wool			120.0–174.0	5.0–10.9

Table 7.
Properties of several natural fibres and commonly used synthetic fibres.

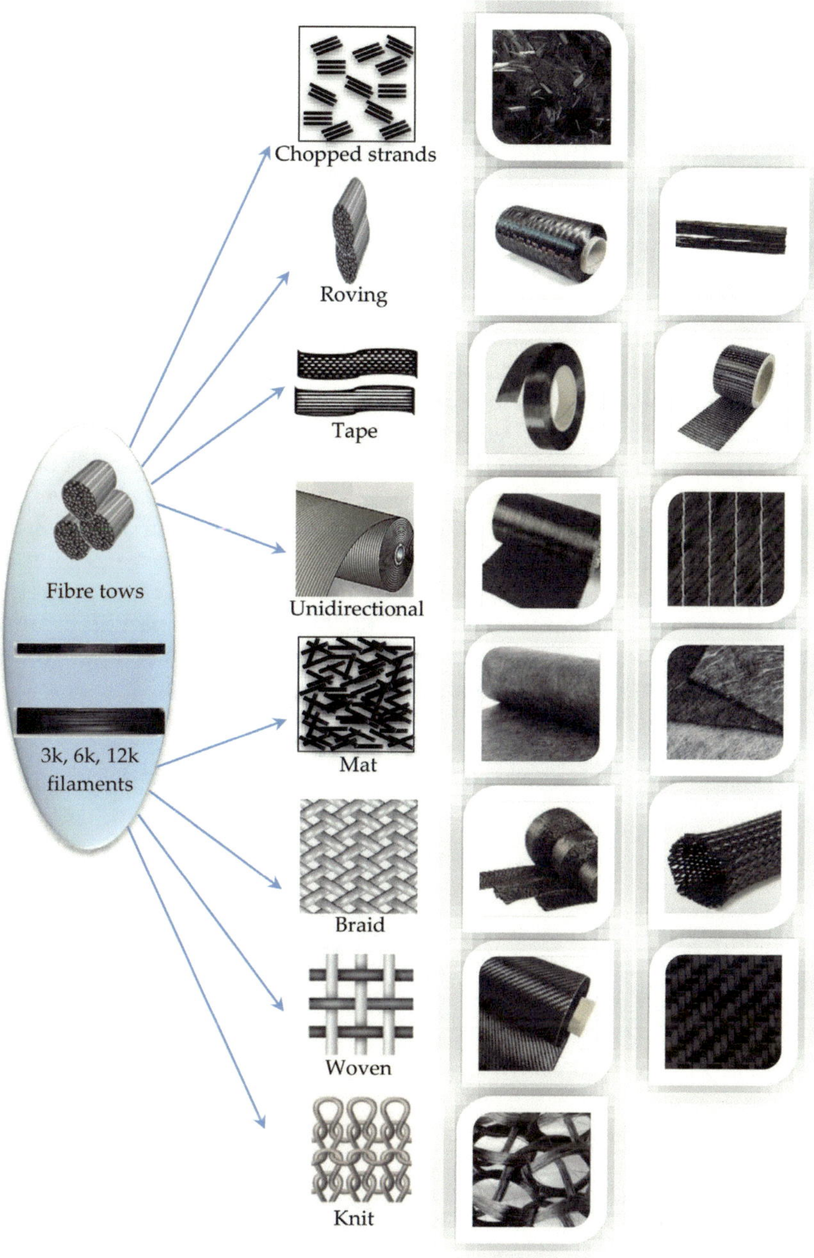

Figure 1.
Various fibre forms.

Increasingly, the fibres have replaced parts formerly made of steel. The fibres used in composite materials appear at different forms and scales as shown in **Figure 1**.

4. Composite manufacturing techniques

There are several methods for fabricating composite materials. The selection of a method for a part will depend on the materials, the part design, the performance, and the end-use or application.

4.1 Open contact moulding

Hand lay-up is an open contact moulding technique for fabricating composite materials. Resins are impregnated by the hand into fibres which are in the form of woven, knitted, stitched, or bonded fabrics. In this technique, the mould is first treated with mould release, dry fibres or dry fabrics are laid on a mould, and liquid resin is then poured and spread onto the fibre beds [36]. This is usually accomplished by rollers or brushes, with an increasing use of nip-roller-type impregnators for forcing resin into the fabrics by means of rotating rollers and a bath of resin. A roller or brush is used to wet the fibres and remove air trapped into the lay-ups. A few layers of fibres are wetted, and laminates are left to cure under standard atmospheric conditions. After these layers are cured, more layers are added, as shown in **Figure 2**.

Spray-up is also an open-mould application technique for composite. The spray lay-up technique is considered an extension of the hand lay-up method. In this process, the mould is first treated with mould release. If a gel coat is used, it is sprayed into the mould at a certain thickness after the mould release has been applied. The gel coat then is cured, and the mould is ready for process. The fibre and catalysed resin at a viscosity of 500–1000 cps are sprayed into the mould using a chopper spray gun. The gun chops continuous fibre tow into short-fibre bundle lengths and then blows the short fibres directly into the sprayed resin stream so that both materials are applied simultaneously on the surface of the mould, as shown in **Figure 3**. In the final steps of the spray-up process, the workers compact the laminate by hand with rollers. The composite part is then cured, cooled, and removed from the mould [37, 38].

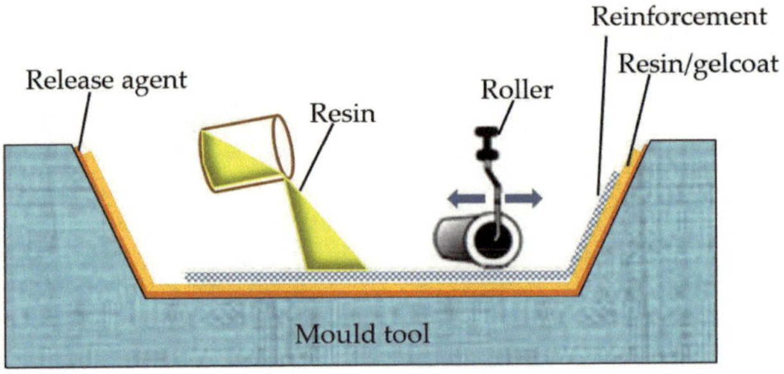

Figure 2.
Hand lay-up process.

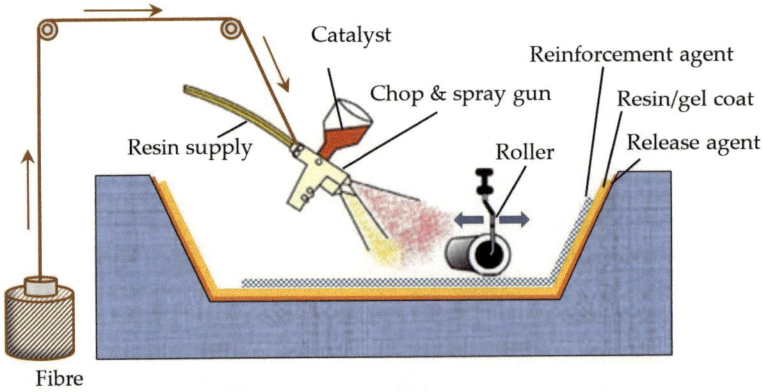

Figure 3.
The schematic of the spray lay-up process.

Hand lay-up and spray-up methods are often used in tandem to reduce labour cost. This is a common process for making glass fibre composite products such as bathtubs, boat hulls and decks, fenders, RV components, shower stalls, spas, truck cabs, and other relatively large and noncomplex shapes.

4.2 Resin infusion processes

With the ever-increasing demand for faster production rates, the industry has used alternative fabrication processes to replace hand lay-up as well as encouraged fabricators to automate those processes wherever possible.

Resin transfer moulding (RTM), sometimes referred to as liquid moulding, is a fairly simple process. In this technique, the mould is first treated with mould release. The dry reinforcement, typically a preform, is then placed into the mould and the mould is closed. Low viscosity resin and catalyst are metered and mixed and then pumped into the mould under low-to-moderate pressure through injection ports, following predesigned paths through the preform. Low-viscosity resin is used in RTM technique to ensure that the resin permeates through the preform quickly and thoroughly before gel and cure, especially with thick composite parts.

Reaction injection moulding (RIM) injects a rapid cure resin and a catalyst into the mould in two separate streams. Mixing and chemical reaction occur in the mould instead of in a dispensing head. Automotive industry suppliers have combined structural RIM (SRIM) with rapid preforming methods to fabricate structural parts that do not require a class A finish. **Figure 4** describes the schematic of the RTM process [39, 40].

Representing the fastest-growing moulding technology is vacuum-assisted resin transfer moulding (VARTM), as shown in **Figure 5**. The difference between VARTM and RTM is that in VARTM, resin is drawn into a preform use a vacuum only, rather than pumped in under pressure as RTM. VARTM technique does not require high heat or pressure. VARTM usually operates with low-cost tooling, making it possible to inexpensively produce large, complex parts in one shot [41–43].

Resin film infusion (RFI) is a hybrid process in which a dry preform is placed in a mould on top of a layer, or interleaved with multiple layers, of high-viscosity resin film. Under applied heat, vacuum, and pressure, the resin liquefies and is drawn into the preform, resulting in uniform resin distribution, even with high-viscosity, toughened resins, because of the short flow distance. Using the resin infusion techniques, the fibre volumes can be up to 70%, and automated controls ensure low voids and consistent preform reproduction, without the need for trimming. Resin infusion has found significant application in boatbuilding. This method has been

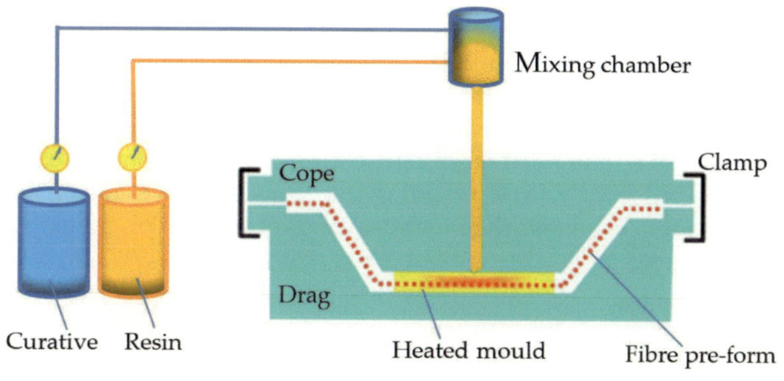

Figure 4.
The schematic of the RTM process.

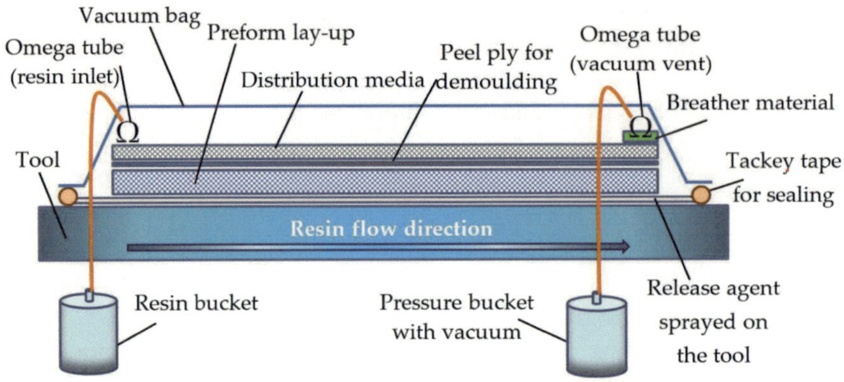

Figure 5.
The schematic of the VARTM process.

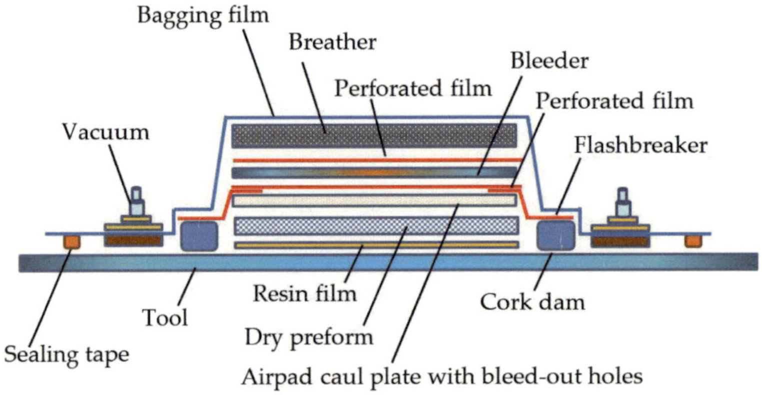

Figure 6.
The schematic of the resin film infusion process.

employed by The Boeing Co. (Chicago, IL, USA) and NASA, as well as small fabricating firms, to produce aerospace-quality laminates without an autoclave [36, 44]. **Figure 6** presents the schematic of the resin film infusion process.

4.3 Compression moulding

Compression moulding is a precise and potentially rapid process for producing high-quality composite parts in a wide range of volumes. The material is manually or robotically placed in the mould. The mould halves are closed, and pressure is applied using hydraulic presses. Cycle time ranges depending on the part size and thickness. This process produces high-strength, complex parts in a wide variety of sizes. The composites are commonly processed by compression moulding and include thermosetting prepregs, fibre-reinforced thermoplastic, moulding compounds such as sheet moulding compound (SMC), bulk moulding compounds (BMC), and chopped thermoplastic tapes. **Figure** 7 shows the schematic of the compression moulding process.

4.4 Injection moulding

Injection moulding is a closed process as shown in **Figure 8**. This is fast, high-volume, low-pressure, and most commonly used for filled thermoplastics, such as nylon with chopped glass fibre. The injection-moulding process has been in use for

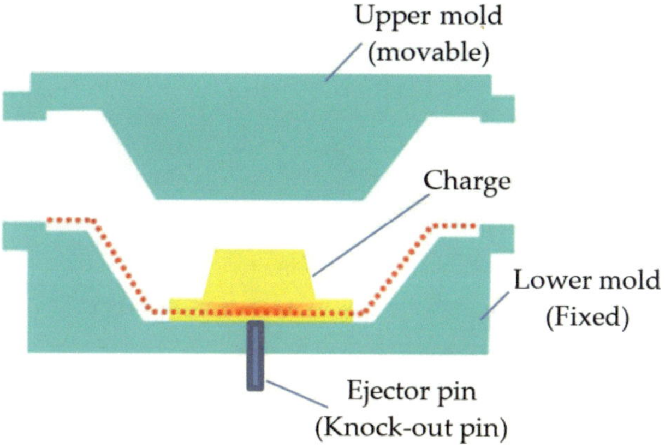

Figure 7.
The schematic of compression moulding process.

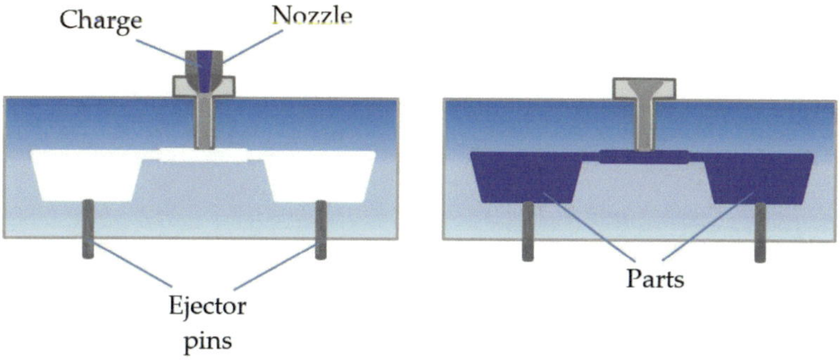

Figure 8.
Simplified diagram of moulding process.

nearly 150 years. Reciprocating screw injection-moulding machines were introduced in the 1960s and are still used today [45]. Injection speeds are typically one to a few seconds, and many parts can be produced per hour in some multiple cavity moulds.

4.5 Filament winding

Filament winding is a continuous fabrication method that can be highly automated and repeatable, with relatively low material costs as shown in **Figure 9**. A long, cylindrical tool called a mandrel is suspended horizontally between end supports. Dry fibres are run through a bath of resin to be wetted. The fibre application instrument moves back and forth along the length of a rotating mandrel with the traverse carriage, placing fibre onto the tool in a predetermined configuration. Computer-controlled filament-winding machines are used to arrange the axes of motion [46–48]. Filament winding is one example of aerospace composite materials.

4.6 Pultrusion process

Composite pultrusion is a processing method for producing continuous lengths of fibre-reinforced polymer structural shapes with constant cross-sections. This is a continuous fabrication method that can be highly automated. In this process, a continuous bundle of dry fibre is pulled through a heated resin-wetting station.

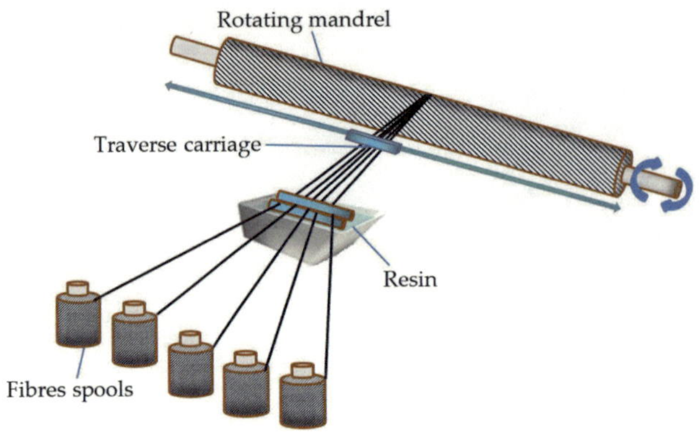

Figure 9.
The schematic of the filament winding process.

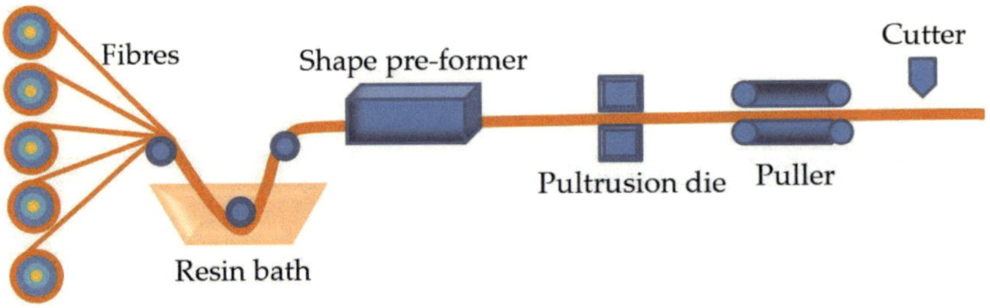

Figure 10.
The schematic of the pultrusion process.

The wetted bundle is pulled into heated dies, and the cross-sectional shape of the pulled fibre is formed by these dies. The resin is cured, and the composites are formed. Parts are then made by slicing the long-cured piece. This process is limited to straight parts with a constant cross-section, such as I-beams, T-beams, or frame sections and ladder rails. **Figure 10** shows the schematic of the pultrusion process [49, 50]. Pultrusion is used in the manufacture of linear components such as ladders and mouldings.

4.7 Automated fibre placement

Automated fibre placement (AFP) is one of the most advanced methods for fabricating and manufacturing of composite materials as presented in **Figure 11**. This method is used almost exclusively with continuous fibre-reinforced tape. A robot is utilised to place fibre-reinforced tape and build a structure one ply (layer) at a time. A band of material comprised of multiple narrow strips of tape (tows) is placed where these tows are commonly 0.125 and 0.25 inches wide. The use of robotics gives the operator active control over all the processes critical variables, making the process highly controllable and repeatable. This method allows the fabrication of highly customised parts as each ply can be placed at different angles to best carry the required loads [51, 52].

Advantages of fibre placement are processing speed and reduced material scrap and labour costs. Often, the process is utilised to fabricate large thermoset parts with complex shapes. Similar to ATP process, automated tape laying (ATL) is an even speedier automated process in which prepreg tape, rather than single tows, is laid down continuously to form parts.

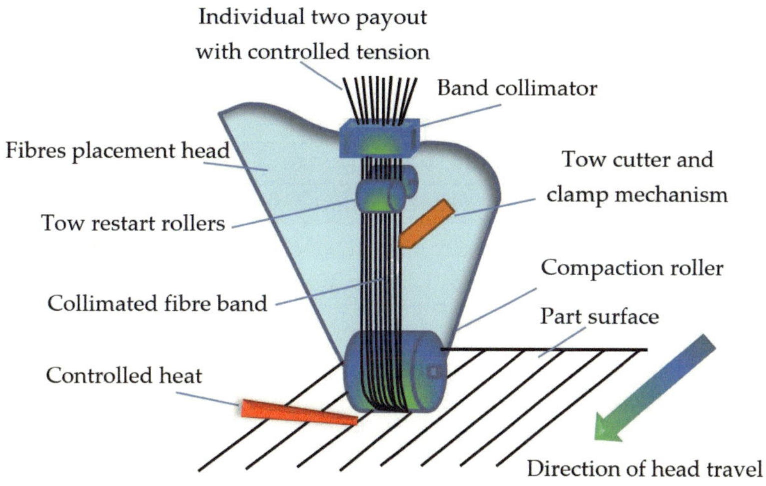

Figure 11.
The schematic of the automated fibre placement process.

4.8 Additive manufacturing

Additive manufacturing is also known as 3D printing technique. Additive manufacturing is a step change in the development of rapid prototyping concepts that were introduced more than 20 years ago. This is a process for making a solid object from a three-dimensional digital model, typically by laying down many successive thin layers of a material. Manufacturing a composite structure with a single nozzle uses polymer composite filament and contains polymer and additives such as rubber microspheres, particles of glass or carbon fibre, wood flour, etc. as shown in **Figure 12**. This more recent form of composite part production grew out of efforts to reduce the costs in the design-to-prototype phase of product development, taking aim particularly at the material-, labour-, and time-intensive area of toolmaking [53–56].

4.9 Composite properties

The polymer composite materials are lightweight, which increases the fuel efficiency of vehicles manufactured from composites and gives them structural

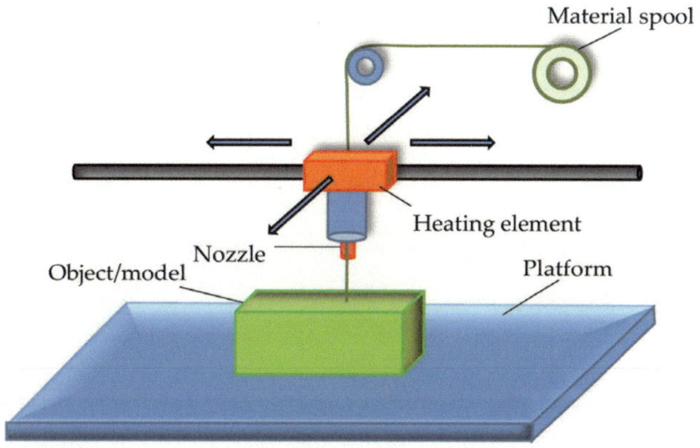

Figure 12.
The schematic of the 3D printing process for polymer composites.

Composite	Density (g/cm^3)	Elongation (%)	Tensile strength (MPa)	Young's modulus (GPa)
ABS + 30% glass fibre	—	2.0	60.0	9.0
Acetal copolymer + 30% glass fibre	—	3.0	110.0	9.5
Epoxy + 40–60% carbon fibre	1.15–2.25	0.4–11.0	4.6–3220.0	2.6–520.0
Epoxy + 45% flax yarn-aligned	—	—	133.0	28.0
Epoxy + 40% glass fibre	—	—	—	—
Epoxy + 52% jute fibre	—	—	216.0	31.0
Epoxy + 52% kevlar fibre	—	—	434.0	28.2
PEEK + 62% carbon fibre	1.60	—	750.0	50.0
PEEK + kevlar fibre	1.31–1.50	1.0–6.0	75.0–193.0	4.0–20.7
Nylon 66 + 25–30% carbon fibre	1.20–1.57	0.90–4.0	193.0–261.0	16.0–33.1
Nylon 6 + 40% glass fibre	1.45	2.0–3.0	235.0	12.9
Nylon + kevlar fibre	1.16	4.0	110.0	9.0
PLA + 30% abaca fibre	—	—	74.0	8.0
PLA + 20% bamboo fibre	—	—	90.0	1.8
PLA + 30% flax fibre	—	—	53.0–100.0	8.0
PF + E glass fibre	1.5–1.65	—	85.0–330.0	5.0–17.0
Polycarbonate + 5–40% carbon fibre	1.15–1.43	0.9–118.0	46.0–186.0	2.1–25.5
Polycarbonate + 30–40% glass fibre	1.44–1.52	4.0	107.0–159.0	10.0–11.6
Polycarbonate-ABS + 30% glass fibre	1.29	—	82.7	—
Polyimide + 20–30% carbon fibre	1.38–1.68	0.8–5.5	36.5–241.0	4.5–29.0
Polyimide + glass fibre	—	2.0	150.0	12.0
PP + 30% carbon fibre	1.07	1.0	117.0	16.2
PP + 30% cotton fibre	—	—	58.5	4.1
PP + 20% glass-chopped strand mat	—	—	77.0	5.4
PP + 20% glass fibre	1.03	3.0–4.0	100.0	4.3
PP + 40% glass fibre	1.22	2.0	127.0	7.6
PP-MAgPP + 40% hemp fibre	—	—	52.0	4.0
UPE + 35% jute fibre	—	—	50.0	8.0
UPE + 47% glass fibre	—	—	201.0	13.0
Vinylester + carbon fibre	1.50–1.65	1.4	900.0–1200.0	136.0
Vinylester + 24% flax fibre	—	—	248.0	24.0
Vinylester + 59% glass fibre	—	—	483.0	33.0
Vinylester + kevlar fibre	1.35	—	500.0	40.0

Table 8.
Properties of several fibre-reinforced polymer composites.

stability. In addition, they offer a high strength-to-weight ratio and increased heat resistance. Composites have very different properties and applications depending on the type of matrix, reinforcement, ratio between them, formulations, processing etc. The bonding strength between fibre and polymer matrix in the composite is considered one of the major factors in order to obtain superior fibre reinforcement polymer composite properties. Typical properties of several polymer composites are presented in **Table 8** [57–71].

The growth of the composites market can be attributed to increased uses in the aerospace, defence, and transportation applications. The global composite materials market is expected to reach an estimated $40.2 billion by 2024, and it is forecasted to grow at a CAGR of 3.3% from 2019 to 2024. The global composite product market is expected to reach an estimated $114.7 billion by 2024 [72].

5. Composite applications

The most widely used form of fibre-reinforced polymer is a laminar structure, made by stacking and bonding thin layers of fibre and polymer until the desired thickness is obtained. By changing the fibre orientation among layers in the laminate structures, a specified level of anisotropy in composite properties can be achieved. Composites offer many benefits such as corrosion resistance, light weight, strength, lower material costs, improved productivity, design flexibility, and durability. Therefore, the wide range of industries uses composite materials and some of their common applications [3, 15].

5.1 Aerospace

The major original equipment manufacturers (OEMs) such as Airbus and Boeing have shown the potential of using composite materials for large-scale applications in aviation. NASA is continually looking to composite manufacturers for innovative approaches and space solutions for rockets and other spacecrafts. Composites with thermoset are being specified for bulkheads, fuselages, wings, and other applications in commercial, civilian, and military aerospace applications. There are several other applications of composites in the areas such as air-foil surfaces, antenna structures, compressor blades, engine bay doors, fan blades, flywheels, helicopter transmission structures, jet engines, radar, rocket engines, solar reflectors, satellite structures, turbine blades, turbine shafts, rotor shafts in helicopters, wing box structures, etc. [3, 15, 26, 37]

5.2 Appliance/business

Composite materials offer flexibility in design and processing; therefore composite materials can be used as alternatives for metal alloys in appliances. Unlike most other industries, trends within the appliance segment move quite quickly. In addition, design and function are subject to both technology advancements and changing consumer taste. Composite materials are being used in appliance and business equipment such as equipment panels, frames, handles and trims in appliances, power tools, and many other applications. Composites are being utilised for the appliance industry in dishwashers, dryers, freezers, ovens, ranges, refrigerators, and washers. The components in the equipment that were utilised composites include consoles, control panels, handles, kick plates, knobs, motor housings, shelf brackets, side trims, vent trims, and many others [3, 73].

5.3 Architecture

With their aesthetic qualities, functionality, and versatility, the composite materials are becoming the material of choice for architectural applications.

Composite materials allow architects to create designs that are impractical or impossible with traditional materials, improve thermal performance and energy efficiency of building materials, and meet building code requirements. Composite materials also offer design flexibility and can be moulded into complex shapes. They can be corrugated, curved, ribbed, or contoured in a variety of ways with varying thickness. Further, a traditional look such as copper, chrome or gold, marble, and stone can be achieved at a fraction of the cost using composite materials. Therefore, the architecture community is experiencing substantial growth in the understanding and use of composites in commercial and residential buildings [15].

5.4 Automotive and transportation

The automotive industry is no stranger to composites. This is one of the largest markets for composite materials. Weight reduction is the greatest advantage of composite material usage. A lower-weight vehicle or truck is more fuel-efficient because it requires less fuel to propel itself forward. In addition to enabling ground breaking vehicle designs, composites help make vehicles lighter and more fuel efficient. The composite materials are used in bearing materials, bodies, connecting rod, crankshafts, cylinder, engines, piston, etc. While fibre-reinforced polymers such as CFRP in cars get most of the attention, composites also play a big role in increasing fuel efficiency in trucks and transport systems. A number of US state Departments of Transportation are also using composite to reinforce the bridges those trucks travel on [3, 26, 37].

5.5 Construction and infrastructure

Construction is one of the largest markets for composites globally. The composites can be made to have a very high strength and ideal construction materials. Thermoset composites are replacing many traditional materials for home and offices' architectural components including doors, fixtures, moulding, roofing, shower stalls, swimming pools, vanity sinks, wall panels, and window frames. Composites are used all over the world to help construct and repair a wide variety of infrastructure applications, from buildings and bridges to roads, railways, and pilings [3, 74].

5.6 Corrosive environments

Products made from composite materials provide long-term resistance to severe chemical conditions and temperature environments. Composites are often the material of choice for applications in chemical handling applications, corrosive environments, outdoor exposure, and other severe environments such as chemical processing plants, oil and gas refineries, pulp and paper converting, and water treatment facilities. Common applications include cabinets, ducts, fans, grating, hoods, pumps, and tanks [3, 37, 73]. Fibre-reinforced polymer composite pipes are used for everything from sewer upgrades and wastewater projects to desalination, oil, and gas applications. When corrosion becomes a problem with pipes made with traditional materials, fibre-reinforced polymer is a solution [3, 73].

5.7 Electrical

With the rapid growth of the electronics industry, and with strong dielectric properties including arc and track resistance, the composite materials are finding more and more in electronic applications. With strong dielectric properties including arc and track resistance, thermoset components include. Applications and components include arc chutes, arc shields, bus supports and lighting components, circuit breakers, control system components, metering devices, microwave antennas, motor controls, standoff insulators, standoffs and pole line hardware and printed wiring boards, substation equipment, switchgear, terminal blocks, and terminal boards [3, 75].

5.8 Energy

Material technology has grown from the early days of glass fibres as major reinforcements for composite material to carbon fibres which are lighter and stronger. The advancements in composites, particularly those from the US Department of Energy, are redefining the energy industry. Composites help enable the use of wind and solar power and improve the efficiency of traditional energy suppliers. Composite materials offer wind manufacturers strength and flexibility in processing with the added benefit of a lightweight components and products [3, 76]. The wind industry has set installation records over the last couple years. According to the Global Wind Energy Council, the trend for this industry may continue with global wind capacity predicted to double in the next few years. Composites play a vital role in the manufacture of structures such as wind turbine blades [3, 77].

5.9 Marine

Just like in the other engineering areas, the main struggle of naval architecture is to achieve a structure as light as possible. The marine industry uses composites to help make hulls lighter and more damage-resistant. With their corrosion resistance and light-weighting attributes, marine composite applications include boat hulls, bulkheads, deck, mast, propeller, and other components for military, commercial, and recreational boats and ships. Composites can be found in many more areas of a maritime vessel, including interior mouldings and furniture on super yachts [3, 78, 79].

5.10 Sports and recreation

The fibre-reinforced composite materials possess some excellent characteristics, including easy moulding, high elastic modulus, high strength, light in weight, good corrosion resistance, and so on. Therefore, fibre-reinforced composite materials have extensive applications in production the manufacturing of sports equipment. From bicycle frames, bobsleds fishing poles, football helmets, hockey sticks, horizontal bars, jumping board, kayaks, parallel bars, props, tennis rackets, to rowing, carbon fibres, and fibreglass composite materials help athletes reach their highest performance capabilities and provide durable and lightweight equipment [3, 80].

6. Summary

Composites have many advantages; a wide range of material combinations can be used in composites, which allows for design flexibility. The composites also can be easily moulded into complicated shapes. The materials can be custom tailored to fit unique specifications. Composites are light in weight compared to most woods and metals and lower density as compared to many metals. They are stronger than some other materials. The materials resist damage from weather and harsh chemicals. Composites have a long service life and require little maintenance. Due to the wide variety of available reinforcement, matrix, and their forms, manufacturing processes, and each resulting in their own characteristic composite products, the design possibilities for composite products are numerous. Therefore, a composite and its manufacturing process can be chosen to best fit the developing rural societies in which the products will be made and applied. Composite materials' research continues. The areas of interest are nanomaterials—materials with extremely small molecular structures and bio-based polymers. To facilitate the advantages of the composites, several aspects must be considered: (a) concept development, (b) material selection and formulation, (c) material design, (d) product manufacturing, (e) market, and (f) regulations.

Acknowledgements

The author acknowledges Mrs. Marian Parslow and Mrs. Laura Parslow for helping in editing of the chapter.

Author details

Tri-Dung Ngo
Biomass Conversion and Processing Technologies, InnoTech Alberta (Formerly Alberta Research Council (1921–2010) and Alberta Innovates Technology Futures (2010–2016)), Edmonton, Alberta, Canada

*Address all correspondence to: tridung.ngo@innotechalberta.ca

References

[1] Johnson T. History of composites. The evolution of lightweight composite materials. 2018. Available from: https://www.thoughtco.com/history-of-composites-820404

[2] Available from: https://www.mar-bal.com/language/en/applications/history-of-composites/

[3] Available from: http://compositeslab.com/composites-101/history-of-composites/

[4] Yuanjian T, Isaac DH. Impact and fatigue behaviour of hemp fibre composites. Composites Science and Technology. 2007;**67**:3300-3307. DOI: 10.1016/j.compscitech.2007.03.039

[5] Sanjay MR, Arpitha GR, Naik LL, Gopalakrishna K, Yogesha B. Applications of natural fibers and its composites: An overview. Natural Resources. 2016;**7**:108-114. DOI: 10.4236/nr.2016.730

[6] Ho MP, Wang H, Lee JH, Ho CK, Lau KT, Leng JS, et al. Critical factors on manufacturing processes of natural fibre composites. Composites Part B Engineering. 2012;**43**:3549-3562. DOI: 10.1016/j.compositesb.2011.10.001

[7] Sathishkumar TP, Navaneethakrishnan P, Shankar S. Tensile and flexural properties of snake grass natural fiber reinforced isophthallic polyester composites. Composites Science and Technology. 2012;**72**:1183-1190. DOI: 10.1016/j.compscitech.2012.04.001

[8] Kalpakjian S, Schmid SR. Manufacturing Engineering and Technology. International Edition. 4th ed. Prentice Hall, Inc.; 2001. ISBN 0-13-017440-8

[9] Cantwell WJ, Morton J. The impact resistance of composite materials—A review. Composites. 1991;**22**(5):347-362. DOI: 10.1016/0010-4361(91)90549-V

[10] Williams J. The Science and Technology of Composite Materials. Australian Academy of Science; 2019. Available from: https://www.science.org.au/curious/technology-future/composite-materials

[11] Available from: http://www.modorplastics.com/thermoset-vs-thermoplastics

[12] Available from: http://www.automateddynamics.com/article/composite-basics/thermoset-vs-thermoplastic-composites

[13] Available from: http://www.transparencymarketresearch.com/thermosets-market.html

[14] Available from: http://articles.sae.org/8547/

[15] Available from: http://compositesmanufacturingmagazine.com/2012/07/exponential-growth-expected-thermoplastic-composites/

[16] Brent Strong A. Fundamentals of composites manufacturing. In: Materials, Methods, and Applications. 2nd ed. USA: Society of Manufacturing Engineers; 2008

[17] Available from: http://www.marketsandmarkets.com/Market-Reports/composites-resin-market-66717271.html

[18] Zitzmann NU, Rateitschak-Pluss E, Marinello CP. Treatment of angular bone defects with a composite bone grafting material in combination with a collagen membrane. Journal of Periodontology. 2003;**74**(5):687-694. DOI: 10.1902/jop.2003.74.5.687

[19] Ramshaw JAM, Werkmeister JA, Peters DE. Collagen as a biomaterial.

In: Williams DF, editor. Current Perspectives on Implantable Devices. London: Jai Press Ltd.; 1990. p. 151

[20] Vert M, Li SM, Spenlehauer G, Guerin P. Bioresorbability and biocompatibility of aliphatic polyesters. Journal of Materials Science: Materials in Medicine. 1992;**3**:432-446. DOI: 10.1007/BF00701240

[21] Ngo T-D. Natural fibres for sustainable bio-composites. In: Gunay E, editor. Natural and Artificial Fiber-Reinforced Composites as Renewable Sources. Croatia: IntechOpen; 2018. DOI: 10.5772/intechopen.71012

[22] Aranguren MI, Reboredo MM. Plant-based reinforcements for thermosets: Matrices, processing, and properties. In: Fakirov S, Bhattacharyya D, editors. Handbook of Engineering Biopolymers: Homopolymers, Blends and Composites. Munich: Hanser; 2007. pp. 193-222

[23] Available from: https://www.designerdata.nl/plastics/thermo+plastics/ASA

[24] Available from: http://d2n4wb9orp1vta.cloudfront.net/resources/images/cdn/cms/0409ct_Glassfiber_3.jpg

[25] Lovell DR. Reinforcements, Chapter 2A. In: Hancox NL, editor. Fibre Composite Hybrid Materials. Barking: Elsevier Applied Science; 1981; ISBN: 0-85334-928-2. PU CSH Library

[26] Hancox NL. Fibre Composite Hybrid Materials. London: Applied Science Publishers; 1981. ISBN: 0-85334-928-2

[27] Available from: https://www.energy.gov/articles/top-9-things-you-didn-t-know-about-carbon-fiber

[28] McConnell V. The making of carbon fiber, high performance composites. Composites World. 2009;**17**(1):37-42. Available from: https://www.compositesworld.com/articles/the-making-of-carbon-fiber

[29] Available from: https://www.fose1.plymouth.ac.uk

[30] Asokan P, Firdoous M, Sonal W. Properties and potential of bio fibres, bio binders, and bio composites. Reviews on Advanced Materials Science. 2012;**30**:254-261. ISSN: 1605-8127

[31] Xue LG, Tabil L, Panigrahi S. Chemical treatments of natural fiber for use in natural fiber-reinforced composites: A review. Journal of Polymers and the Environment. 2007;**15**:25-33. DOI: 10.1007/s10924-006-0042-3

[32] Saheb DN, Jog JP. Natural fiber polymer composites: A review. Advances in Polymer Technology. 1991;**18**:351-363. DOI: 10.1002/(SICI)1098-2329(199924)18:4

[33] Mohanty AK, Misra M, Hinrichsen G. Biofibres, biodegradable polymers and biocomposites: An overview. Macromolecular Materials and Engineering. 2000;**276-277**(1):1-24. DOI: 10.1002/(SICI)1439-2054(20000301)276:1

[34] Jawaid M, Abdul Khalil HPS. Cellulosic/synthetic fibre reinforced polymer hybrid composites: A review. Carbohydrate Polymers. 2011;**86**(1):1-18. DOI: 10.1016/j.carbpol.2011.04.043

[35] Latif R, Wakeel S, Khan NZ, Siddiquee AN, Verma SL, Khan ZA. Surface treatments of plant fibers and their effects on mechanical properties of fiber-reinforced composites: A review. Journal of Reinforced Plastics and Composites. 2018;**0**(0):1-16

[36] Available from: https://netcomposites.com/guide/manufacturing/wet-hand-lay-up/

[37] Available from: https://www.compositesworld.com

[38] Available from: https://www.eppcomposites.com/spray-layup-process.html

[39] Kendall KN, Rudd CD, Owen MJ, Middleton V. Characterization of the resin transfer moulding process. Composites Manufacturing. 1992;**3**(4): 235-249. DOI: 10.1016/0956-7143(92)90111-7

[40] Available from: https://en.wikipedia.org/wiki/Transfer_molding

[41] Heider D, Graf A, Fink Bruce K, Gillespie JW Jr. Feedback control of the vacuum-assisted resin transfer molding (VARTM) process. Process Control and Sensors for Manufacturing. 1999;**II** (3589):133-141. DOI: 10.1117/12.339956

[42] Hayward JS, Harris B. The effect of vacuum assistance in resin transfer moulding. Composites Manufacturing. 1990;**1**(3):161-166. DOI: 10.1016/0956-7143(90)90163-Q

[43] Lih TC, Azm AI, Muhammad N. Delamination and surface roughness analyses in drilling hybrid carbon/glass composite. Materials and Manufacturing Processes. 2016;**0**:1-11. DOI: 10.1080/10426914.2015.1103864

[44] Qi B, Raju J, Kruckenberg T, Stanning R. A resin film infusion process for manufacture of advanced composite structures. Composite Structures. 1999;**47**(1–4):471-476

[45] Frizelle WG. Injection molding technology. In: Kutz M, Andrew W, editors. Applied Plastics Engineering Handbook. USA: Elsevier; 2011. p. 2011

[46] Minsch N, Herrmann FH, Gereke T, Nocke A, Cherif C. Analysis of filament winding processes and potential equipment technologies. Procedia CIRP. 2017;**66**:125-130. DOI: 10.1016/j.procir.2017.03.284

[47] Ma Q, Rejab R, Kaige J, Idric M, Harith MN. Filament winding technique, experiment and simulation analysis on tubular structure. In: IOP Conference Series: Materials Science and Engineering 342 012029, iCITES 2018. 2018. DOI: 10.1088/1757-899X/342/1/012029

[48] Tarnopol'skii YM, Peters ST, Beil AI. Filament winding. In: Peters ST, editor. Handbook of Composites. Boston, MA: Springer. DOI: 10.1007/978-1-4615-6389-1-22

[49] Joshi SC. The pultrusion process for polymer matrix composites. In: Advani SG, Hsiao KT, editors. Manufacturing Techniques for Polymer Matrix Composites (PMCs). Cambridge, UK: Woodhead Publishing Limited; 2012. ISBN: 978-0-85709-067-6

[50] Wilson BA. Pultrusion. In: Peters ST, editor. Handbook of Composites. Boston, MA: Springer; 1998. DOI: 10.1007/978-1-4615-6389-1-24

[51] Oromiehie E, Gangadhara P, Compston P, Rajan G. Automated fibre placement based composite structures: Review on the defects, impacts and inspections techniques. Composite Structures. 2019;**110987**:224. DOI: 10.1016/j.compstruct.2019.110987

[52] Marsh G. Automating aerospace composites production with fibre placement. 2011. Available from: https://www.materialstoday.com/carbon-fiber/features/automating-aerospace-composites-production-with/

[53] Bryll K, Piesowicz E, Szymański P, Ślączka W, Pijanowski M. Polymer composite manufacturing by FDM 3D printing technology. In: MATEC Web of Conferences 237, 02006. 2018. DOI: 10.1051/matecconf/2018237020

[54] Dudek P. FDM 3D printing technology in manufacturing composite elements. Archives of Metallurgy and Materials. 2013;**58**(4):1415-1418

[55] Apparatus and method for creating three-dimensional objects. Patent US 5121329 A. 1992

[56] Ning F, Cong W, Qiu J, Wei J, Wang S. Additive manufacturing of carbon fiber reinforced thermoplastic composites using fused deposition modeling. Composites Part B: Engineering. 2015;**80**:369-378

[57] Available from: http://www.matweb.com/reference/tensilestrength.aspx

[58] Van de Weyenberg I, Ivens J, De Coster A, Kino B, Baetens E, Verpoest I. Influence of processing and chemical treatment of flax fibres on their composites. Composites Science and Technology. 2003;**63**(9):1241-1246. DOI: 10.1016/S0266-3538(03)00093-9

[59] Ozsoy N, Mimaroğlu A, Ozsoy M, Ozsoy MI. Comparison of mechanical behaviour of carbon and glass fiber reinforced epoxy composites. Acta Physica Polonica A. 2015;**127**(4): 1031-1034. DOI: 10.12693/APhysPolA.127.1032

[60] Rodriguez E, Petrucci R, Puglia D, Kenny JM, Vazquez A. Characterization of composites based on natural and glass fibers obtained by vacuum infusion. Journal of Composite Materials. 2005; **39**(3):265-282. DOI: 10.1177/0021998305046450

[61] Seena J, Sreekala MS, Oommen Z, Koshy P, Sabu T. A comparison of the mechanical properties of phenol formaldehyde composites reinforced with banana fibres and glass fibres. Composites Science and Technology. 2002;**62**(14):1857-1868. DOI: 10.1016/S0266-3538(02)00098-2

[62] Oksman K. Mechanical properties of natural fibre mat reinforced thermoplastic. Applied Composite Materials. 2000;**7**(5–6):403-414. DOI: 10.1023/A:1026546426764

[63] Sain M, Suhara P, Law S, Bouilloux A. Interface modification and mechanical properties of natural fiber-polyolefin composite products. Journal of Reinforced Plastics and Composites. 2005;**24**(2):121-130. DOI: 10.1177/0731684405041717

[64] Oksman K, Skrifvars M, Selin JF. Natural fibres as reinforcement in poly-lactic acid (PLA) composites. Composites Science and Technology. 2003;**63**:1317-1324. DOI: 10.1016/S0266-3538(03)00103-9

[65] Bodros E, Pillin I, Montrelay N, Baley C. Could biopolymers reinforced by randomly scattered flax fibre be used in structural applications? Composites Science and Technology. 2007;**67**(3): 462-470. DOI: 10.1016/j.compscitech.2006.08.024

[66] Goutianos S, Peijs T, Nystrom B, Skrifvars M. Development of flax fibre-based textile reinforcements for composite applications. Applied Composite Materials. 2006;**13**(4): 199-215. DOI: 10.1007/s10443-006-9010-2

[67] Available from: http://www.matweb.com/search/datasheettext.aspx?matguid=39e40851fc164b6c9bda29d798bf3726

[68] Sanjeevamurthy GC, Srinivas GR. Sisal/coconut coir natural fibers-epoxy composites: Water absorption and mechanical properties. International Journal of Engineering and Innovative Technology. 2012;**2**(3):166-170. ISSN: 2277-3754

[69] Sawpan MA, Pickering KL, Fernyhough A. Improvement of mechanical performance of industrial hemp fiber reinforced polylactide biocomposites. Composites Part A: Applied Science and Manufacturing. 2011;**42**(3):310-319. DOI: 10.1016/j.compositesa.2010.12.004

[70] Kalia S, Kaith BS, Kaur I. Pretreatments of natural fibers and their application as reinforcing material in polymer composites—A review. Polymer Engineering & Science. 2009; **49**(7):1253-1272. DOI: 10.1002/ pen.21328

[71] Chehroudi B. Composite Materials and Their Uses in Cars. Part I: What is a Composite Material? Advanced Technology Consultants; 2017. Available from: https://advtechconsultants.com

[72] Available from: www.plasticomp. com

[73] Hu N. Composites and Their Applications. Croatia: IntechOpen; 2012. DOI: 10.5772/3353

[74] Erofeev V. Frame construction composites for buildings and structures in aggressive environments. Procedia Engineering. 2016;**165**:1444-1447

[75] Chung DDL. Composite materials for electrical applications. In: Composite Materials. Engineering Materials and Processes. London: Springer; 2003

[76] Available from: https://www.windpowerengineering.com/blade-materials-manufacturing-changing-keep-larger-turbines/

[77] Calvário M, Sutherland LS, Guedes Soares C. A review of the applications composite materials in wave and tidal energy devices. In: Gunay E, editor. Natural and Artificial Fiber-Reinforced Composites as Renewable Sources. Developments in Maritime Transportation and Harvesting of Sea Resources: Proceedings of the 17th International Congress of the International Maritime Association of the Mediterranean (IMAM 2017), October 9–11, 2017, Lisbon, Portugal. CRC Press; IntechOpen, Croatia. 2018. DOI: 10.5772/intechopen.71012

[78] Pemberton R, Summerscales J, Graham-Jones J. Marine Composites: Design and Performance. Cambridge, UK: Woodhead Publishing; 2018. DOI: 10.1016/C2016-0-00710-6

[79] Neşer G. Polymer based composites in marine use: History and future trends. Procedia Engineering. 2017;**194**: 19-24

[80] Sun GH, Wang JJ. The applied research of fiber reinforced composites materials in sports equipments. Advanced Materials Research. 2012;**485**: 506-509

Chapter 2

Introduction of Natural Fiber Composite Using in Paperboard Industry

N. Venkatachalam

Abstract

India is one of the largest countries having agriculture as the main profession for many people. A huge amount of natural fibers are present in the plants of agricultural land and forest. As these natural fibers have high potential due to their biodegradable nature, new ways must be found for the utilization of these natural fibers in industries. Nowadays, the paperboard producing ventures are confronted with two main problems, namely environmental pollution and insufficiency of raw materials. These ventures necessitate the spending of a lot of energy and money for diminishing the level of contamination in paperboard effluents. One method of diminishing the contamination in paperboard industries is the utilization of natural fiber composites. For the last two decades, there has been a lack of raw materials in paperboard industries. Consequently, it is important to find new resources. Presently, many researches focus on the development of natural fiber-reinforced composites for paperboard. The natural fiber composites could be used to solve the above two issues.

Keywords: chemically treated fibers, natural fiber composites, sawdust, properties, paperboard industry

1. Introduction

India is one of the biggest agricultural countries with agriculture as the main profession for many people. A large quantity and variety of natural fibers are available from the plants in agricultural land and forests. New ways have to be found for making use of these natural fibers in various industrial products. Presently, paperboard manufacturing industries face two major challenges, namely environmental pollution and deficiency of raw materials. These manufacturing industries have to put in more efforts and money for reducing the level of pollution in paperboard effluents. One way of reducing pollution in paper industries is the use of natural fiber composites. In the past two decades, worldwide raw material deficiency has been occurring in paperboard raw materials. Hence, it is necessary to go for alternative materials. Much interest is shown in the extensive research on using natural fiber reinforced composites in paperboard industry. The use of natural fiber composites is the solution for the above two problems.

In India, endeavors have been made to fill the gap between demand and supply of paperboard by utilizing different agricultural products and weeds. The natural fiber in reinforced polymer composite is considered as a substitute for synthetic fiber. The natural fibers have high specific modulus and moderate strength. They are completely combustible, affordable, lightweight, non-toxic as well as easy to recycle. The practical attributes necessary for paperboard are strength, anti-fungus, printability, processability, recyclability and biodegradability. The added advantage of using natural fiber composite in paperboard is that it reduces the global warming. The natural fiber composites solve the environmental effect to a great extent by zero emission of CO_2 to the atmosphere. The natural fibers can be produced at a relatively low cost and low specific weight compared to synthetic fibers.

1.1 Natural fibers

Natural fibers can be classified into three main categories: vegetable fibers, animal fibers and mineral fibers. Vegetable fibers are composed primarily of cellulose, hemicellulose and lignin, with the balance being made up of pectin, water soluble compounds, wax, inorganic and non-flammable substances which are generally referred to as ash. The structure, microfibrillar angle, cell dimensions, defects and the chemical composition of fibers are the most important variables that determine the overall properties of the fibers. Plant fibers will exhibit high ductility if the microfibrils have a spiral orientation to the fiber axis. The properties of plant fibers depend on the type and age of plant, type of soil, climate conditions, the extraction method used, the fiber structure, microfibrillar angle, fiber (cell) dimensions and chemical composition.

The contents of natural fibers are cellulose, hemicellulose, lignin, ash and moisture. The cellulose content has an important influence on the mechanical properties of fiber such as tensile strength, Young's modulus and strain-to-failure.

The physical properties of natural fibers like density and diameter are the most important properties to make the light weight composites. The mechanical performance of the fiber-reinforced composites is mainly the function of the fiber dispersion, fiber-matrix compatibility and aspect ratio of the reinforcement. The tensile properties such as tensile strength, modulus and strain-to-failure of natural fibers play an important role in deciding the properties of polymer composites.

1.2 Chemically treated fibers

Natural fibers are noted for their hydrophilic nature due to the high quantity of hydroxyl groups gathered in cellulose. One way of improving the interfacial bonding of the fibers with matrix is fiber surface modification. The cellulose of natural fiber contains hydrophilic nature, whereas lignin contains hydrophobic nature. Therefore, they are subject to modification. The hydroxyl groups may be required in the hydrogen bonding within the cellulose molecules, thereby reducing the activity towards the matrix.

The interest towards using natural fiber as reinforcement in composites has increased dramatically and it represents one of the most important uses in paperboard industry. Cellulosic fibers are hygroscopic in nature; moisture absorption can result in the swelling of the fibers which may extend to micro-cracking of the composite and degradation of mechanical properties. This problem can be overcome by treating these fibers with suitable chemicals to decrease the hydroxyl groups which may be required in the hydrogen bonding within the cellulose molecules. The mechanical and chemical bondings at the fiber surface are mainly dependent on the surface morphology and chemical composition of the fibers. Therefore, the microscopic analysis of fiber surface topology and morphology is of extreme importance in fibrous composites. Hence, in the present research, an attempt has been made for heat, alkalization, silane, acetylation, acrylation, permanganate and benzoylation treatments of fiber to modify the physical properties, morphology, crystallinity and thermal stability of *Passiflora foetida* Fiber (PFF). The reinforcing efficiency of natural fiber depends upon the cellulose and its crystallinity.

To identify the influence of treatment on the fibers, chemical composition, the Fourier Transform Infrared (FTIR) spectra is utilized. The crystalline structure of cellulose in PFF is detected by wide angle X-Ray Diffraction (XRD) spectra. Thermal degradation of fibers is analyzed by Thermogravimetric Analysis/ Differential Thermogravimetric analysis (TGA/DTG).

1.3 Sawdust (SW) filler

There are at present million sawmills working in urban region of India. A large quantity of SW is generated in sawmill industry. Fundamentally, this natural SW is utilized as a fuel source as well as a source for furniture item, namely plywood. The3'R concepts, namely "Reduce, Recycle and Reuse" can be applied to filler. SW represents an imperative renewable source of filler as a part of biopolymer composites.

SW is important in light of thc fact that it is reused of ease, low weight and high stiffness. It diminishes shrinkage of composite in the wake of embellishment. Hence, the SW is tried as a filler material in a newly identified PFF with polymer composite and the resulting modified properties of filler are analyzed. The treated SW filler is randomly dispersed in the matrix such a way to obtain a homogeneous and isotropic macroscopic behavior of composite.

1.4 Fiber paperboard properties

Leu et al. [1], Stepanov et al. [2], Gonzalez et al. [3] and Adu and Jolly [4] have pointed out that the paperboard materials are classified into three types based on the source materials used for preparation as shown in **Figure 1**.

Increasing global consumption, modernization and luxury of the paperboard make a need to cut 7.2 billion trees for paper production for packing purpose. Of late, the raw material required for paperboard is 10 million tons per annum and this may be increased to 20 million tons per annum from 2020. Madras Consultancy Group (MCG) has reported that 21% of the world's packaging market is done in India. The non-food material packing by paperboard has increased to 6.3% and this may also be increased to 2.4% per annum for next 5 years. Due to the shortage of raw materials, the paperboard industries are looking for the non-wood plant wastes. This may attract the researchers to investigate the alternate raw materials for paperboard products.

The pulp has been used as a raw material for paperboard industries. During the sulfur emission, the processing of paper may cause air and land pollution. The carbon-positive, water-positive, zero-solid waste disposal, Elemental Chlorine Free (ECF) technology and ozone technologies are being used to reduce toxins in

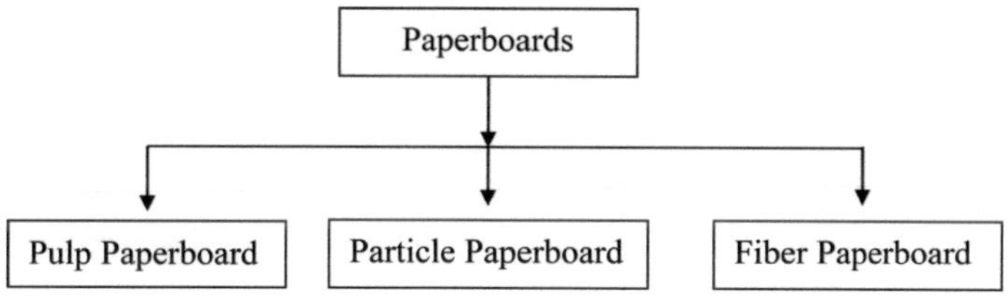

Figure 1.
Types of paperboard materials.

the effluent discharged from paper industries. These technologies help in producing brighter and stronger paper products. By using these above technologies, the pollutants can be kept within the limits as per the Biochemical Oxygen on Demand (BOD) and Chemical Oxygen on Demand (COD) national standards.

Manufacturing industries have to take more efforts and spend much money for reducing the level of pollution in paperboard effluents. For reducing pollution in paper industries, natural fiber composites are used in the present study. The following properties are suitable for paperboard. Paperboard should have higher ductility, modulus, impact strength, durability, fungal and water absorption resistance. It should possess elasto-visco-plastic properties, namely rheological behaviors such as delayed strain recovery, stress relaxation and creep resistance.

Basically, the natural fibers are in the form of small hollow structure. Brindha et al. [5] and Sharma et al. [6] have found that natural fibers with cellulose of 34% and above and less than 30% of lignin are suitable for paperboard production. Runkel's ratio, slenderness ratio and flexibility coefficient are significantly derived indices to determine the suitability of composite material for paperboard making. In composite materials, Runkel's ratio less than 1, slenderness ratio more than 33 and flexibility ratio between 60 and 70 of fibrous materials are good for paperboard making because fibers are more flexible and they would collapse easily and form a paperboard with large bonded area. As per ISO standard 536, the minimum thickness of 0.25 mm and grammage above 224 g/m^2 are required for paperboard.

1.5 Determination of the density of PFFs

To calculate the volume fraction of fibers in composite, it is necessary to determine the density of PFFs. The fiber density is found using pycnometer as per ASTM D 578-89. The fibers are mixed with methyl benzene (known as toluene) immersion liquid. Initially, the fibers are impregnated in methyl benzene for 2 hours to remove the micro bubbles. Then the fibers are dried for 2 days in air tight non-hygroscopic desiccator containing calcium chloride. The density of the toluene (ρ_T) is 0.8669 g/cm^3 at 20°C. Then the fibers are cut into 10 mm and placed in the pycnometer. The density of PFFs is calculated by the expression.

$$\rho_{PFFs} = \left(\frac{m_2 - m_1}{(m_3 - m_1)(m_4 - m_2)}\right) \times \rho_T$$

where m_1 is the mass of the unfilled pycnometer (kg), m_2 is the mass of the pycnometer loaded with cleaved fibers (kg), m_3 is the mass of the pycnometer loaded with methylbenzene (kg) and m_4 is the mass of the pycnometer loaded with hacked fibers and methylbenzene solution (kg). The density of natural fiber is found.

1.6 Properties of polyester

Based on the reports obtained from the supplier's datasheet, the properties of the polyester resin are as listed in **Table 1**.

The density of the composite is low due to the addition of lower density PFFs.

1.7 Determination of the density of SW

The density of SW is determined by Maharani et al. [7]. It is estimated by the addition of SW into a volumetric cylinder of 100 cm^3 (v_o). Itis calculated by the expression.

$$\text{SW density} = (W_a - W_b)/V_o$$

Properties	Values
Density	1.1 g/cm^3
Tensile strength	16–18 MPa
Tensile modulus	0.25–0.4GPa
Flexural strength	30 MPa
Flexural modulus	1.1–1.4 GPa

Table 1.
Properties of polyester resin.

Where w_a is the weight of empty volumetric cylinder while w_b is the combined weight of SW and volumetric cylinder. The density of SW is 1.17 g/cm^3.

1.8 Procedure for v_f to w_f

To achieve the desired volume of fiber and matrix in the composite, the weight of fiber and matrix is to be calculated as follows:

1. To find the density of composite, $\rho_c = \rho_f V_f + \rho_m V_m$

2. To calculate the volume of composite from pattern size

$$V_c = \text{Length} \times \text{width} \times \text{thickness}.$$

3. To find the mass of composite, $m_c = \rho_c \times V_c$

4. To calculate the weight percentage of fiber, $\%w_f = \frac{\rho_f}{\rho_c} \times V_f$

5. To find the weight of the fiber to be added in the mold, $w_f = \%w_f \times m_c$

The present book deals with the mechanical properties, chemical compositions, crystallinity and thermal stability of the newly developed composites. Finally, the results are analyzed to ensure the suitability of natural fiber as an optional fiber-material for paperboard industry.

Author details

N. Venkatachalam
Excel Engineering College, Komarapalayam, Tamilnadu, India

*Address all correspondence to: nvenk@yahoo.co.in

References

[1] Leu S, Yang T, Lo S, Yang T. Optimized material composition to improve the physical and mechanical properties of extruded wood–plastic composites (WPCs). Construction and Building Materials. 2012;**29**:120-127

[2] Stepanov A, Saukkonen E, Piili H. Possibilities of laser processing of paper materials. Physics Procedia. 15th Nordic Laser Materials Processing Conference. Lappeenranta, Finland. 2015;**78**:138-146

[3] Gonzalez VA, Cervantes MJ, Olayo R, Franco HPJ. Effect of fiber surface treatment on the fiber–matrix bond strength of natural fiber reinforced composites. Composites: Part B. 1999;**30**:309-320

[4] Adu C, Jolly M. Developing fiber and mineral based composite materials from paper manufacturing by-products. In: Proceedings of International Conference on Sustainable Design and Manufacturing, SDM 2017. Vol. 68. 2017. pp. 435-444

[5] Brindha D, Vinodhini S, Alarmelumangai K. Fiber dimension and chemical contents of fiber from Passiflora Foetida, L and their suitability in paper production. Science Research Reporter. 2012;**2**(3):210-219

[6] Sharma M, Sharma CL, Kumar YB. Evaluation of fiber characteristics in some weeds of Arunachal Pradesh, India for pulp and paper making. Research Journal of Agriculture and Forestry Sciences. 2013;**1**(3):15-21

[7] Maharani R, Yutaka T, Yajima T, Minoru T. Scrutiny on physical properties of sawdust from tropical commercial wood species: Effects of different mills and sawdust's particle size. Journal of Forestry Research. 2010;**7**(1):20-32

Section 2

Aramid and Carbon Fibers Composites

Chapter 3

Advanced Treatments of Aramid Fibers for Composite Laminates

Mikko Kanerva

Abstract

Aramid fibers form an important group of fibers for composite applications. These applications range through light-weight shell structures, protective structures in ballistic applications such as helmets and various shields, protective clothing, and car tires, for instance. For structural applications, the composites of aramid fibers and high performance resins must form integral and strong parts. Therefore, the fiber-matrix interface places a significant role. Numerous surface treatments and fiber modifications have been applied over the years to adjust aramid fibers. On the way to improve and optimize these interfaces, various test methods have been applied. The recent studies apply microtesting, e.g., in the form of microdroplet tests. Furthermore, the material properties of the resin, fiber, and interface are used to create numerical models. However, the current challenge is to collect statistically reliable data as well as the necessary parameters to validate the simulations on different length scales.

Keywords: aramid, fiber, interface, finite element analysis, microdroplet

1. Introduction

Aramid fibers are a group of tough and strong fibers. The uniting feature of aramid fibers is the synthetic, polymeric base of the fibers. This polymeric base contains essentially a repeating structure of benzene rings connected by amide bonds and interchain hydrogen bonds, and the vast group of these polymers and especially the spun fibers are commonly called aramid fibers. Historically, the commercial fibers were established after discovering the so-called lyotropic liquid crystalline aramids. The first successful spinning trials were run by a famous chemist S. Kwolek while working for the company DuPont. Nowadays, commercial trade names of Kevlar®, Twaron®, and Technora®, are well known, provided by different manufacturers.

2. Applications

Aramid fibers have an inherent yellowish color that often is a sign of well-known applications of these fibers. Aramid fibers are good in light-weight applications because the density of the fibers is relatively low, in the order of 1.35–1.5 g/cm^3. Naturally, the stiffness (modulus) and strength of the fibers finally determine the applications for optimum designs. Indeed, aramid fibers are commercially provided with a range of stiffness and ultimate strength. The fibers that have the highest stiffness are common in structures where the design is dictated by deformation. Various secondary and tertiary structures in aircraft are typical, such as jet engine cowlings, leading edges of wings, and tail planes. It should be noted that it is rather a common tradition to apply fabrics of aramid fibers in combination with other reinforcements such as carbon and glass fiber-based fabrics in laminate lay-ups. If seen important, aramid fibers with Young's modulus matching that of glass fibers can be used. Modulus-matched designs can decrease the amount of internal stresses in composite structures, so that the durability is increased.

The fact is that aramid fibers are tough, and they can extensively absorb energy in different dynamic loading conditions. Indeed, aramid fibers are applied in structures where a high level of energy absorption is required. These fibers are superior to glass and carbon rivals that cannot meet the requirements in shielding constructions. This is typically the critical requirement when a structure is expected to be damaged by an impact during operation. Thus, a damaged structure is considered from the point of view of damage tolerance or damage resistance.

Clearly, aramid fibers are an optimum raw material option for ballistic applications—the examples of products range from composite helmets to protective shields and vests against ammunition. Importantly, by selecting aramid fibers with a lower stiffness, they can well be used in systems with elastomeric bimaterials. The combination with rubbery polymers refers to shielding applications, such as flame- and cut-resistant clothing and gloves. Due to the very high toughness, various kinds of belts and ropes with fiber reinforcement have been developed using aramid fibers for industry and automotive machinery—in these elastomer-fiber applications, glass and carbon fibers are not an option. In car tires, aramid fiber reinforcements form the load-carrying structure of the most high-quality tires.

There are many applications for aramid fibers where they are not much visible. Chopped fiber form is typical reinforcement in clutch plates and brake pads. For the most requiring suburban locations, aramid fiber particles are used as the reinforcing component in cement and road surfacing. Most beautifully, sailboat sails are an application where all the features of aramid fibers are in operation at their full extent: maximum strength, stiffness, and ductility.

After all, it should be remembered that there are other design requirements than the mechanical allowables. Actually, the other requirements can be governing the product development. These requirements on the fiber level can refer to electrical conductivity, especially personalized esthetics, antibacterial activity, and extensive vibration damping. In the event of aramid fibers applied in the system, additional modification of the fibers or their surfaces must be accomplished.

3. Surface characteristics and fiber treatments

Surfaces of aramid fibers can be modified due to various reasons, and the enhancement of the adhesion to composites' matrix resins is not the least emphasized reason. The challenge with the surfaces of aramid fibers is twofold:

I. The adhesion in general between aramid fibers and various polymers is poor [1, 2].

II. The subsurface structure in aramid fibers can lead it to peeling off [3, 4].

The first of the above challenges is of primary importance because the latter one results only in the event of appropriate adhesion.

Therefore, many of the studies concerned with the poor adhesion had the target of improving the adhesion. As a starting point, due to the chemistry of the aramid polymer, the fiber surfaces are highly inert and do not dissolve in almost any practical solvent. Besides, as is typical for spun fibers, the surfaces of the fibers are smooth.

For aramid fibers, the fundamental theories forming the basis of understanding adhesion include [5, 6]:

I. weak atomic interaction, i.e., the van der Waals forces between surfaces;

II. intermediate atomic forces, i.e., ionic and hydrogen bonds;

III. primary, metallic, and covalent bonds;

IV. electrostatic forces; and

V. mechanical interlocking (cohesive resistance).

The formation of the fiber-resin interface is largely affected by the polarity and total surface energy of the fiber surface. Thus, the addition of polar groups has been suggested as a viable solution to increase adhesion. Various oxidation treatments [1] form the basic methodology to modify fiber surfaces, and these procedures cover: (1) gaseous oxidation treatments; (2) solution oxidation treatments; and (3) electrochemical or electrolytic oxidation treatments. These treatments merely modify the fiber surface morphology and might give a change in the surface energy and chemistry as well. In addition to the oxidation treatments, plasma treatments have been applied to aramid fibers. Brown and Mathys [7] applied ammonia and oxygen plasma treatments and reported enhanced performance of laminates, in terms of interlaminar shear strength. Shaker et al. [8] applied radio-frequency (R-F) plasma to modify aramid fibers and reached improved laminate properties.

The application of surface modifications, to allow mechanical interlocking, has been introduced by Lee-Sulivan et al. [9], Palola et al. [10], and Wu et al. [11]. Naturally, the modification of surface roughness also affects the surface chemistry of the fibers. Lin [12] studied the use of bromination and metalation to modify surface roughness and chemistry. Very often effective and strong surface treatments lead to deterioration and lower strength and stiffness of the fibers, although macroscopic laminate properties of a composite might remain on an acceptable level or even on a high level [12].

In detail, the fiber-matrix interface, as a definition, is not entirely a black-and-white plane. Already due to the physical boundary set by the fiber and molecular orientation at the surface, an intermediate phase (interphase) forms with macromolecular polymer resins. Engineering of this interphase, and even its removal, has been reported [13, 14]. The typical surface treatments by solutions, namely sizings of aramid fibers are applications of emulsified solutions, e.g., solutions doped with ethylene oxide and propylene oxide chain fractions on fibers or bundles. Sizings based on aqueous epoxy-piperazine solutions were studied by de Lange et al. [13], and they reported improved adhesion in terms of bundle pull-out force. The identification of an interphase via X-ray photoelectron spectroscopy (XPS) and low-energy ion scattering tends to be difficult due to common elemental species, such as those represented by nitrogen and oxygen bands, detected in the fibers and anticipated interphases. A fiber finish or sizing often includes oily low-molecular weight components, which might lead to a wider interphase formation but eventually, for overly large concentration, lower adhesion [15].

Many of the studies of surface treatments anticipated that the treatments did not result in covalent bonds between the fiber surface and the matrix polymers [15, 16]. Only few works reported the potential of covalent bonds [7, 17].

The optimization of surface treatments is important to gain the desired macroscopic behavior in specific composite laminates. Naturally, the recipe of an optimum treatment depends on the targeted laminate behavior. Due to the typical applications of aramid fibers and their reinforcements, impact tests are frequently used in the evaluation of the interface performance on a laminate scale. Impact resistance and impact damage are complex phenomena. Good interfacial adhesion does not necessarily lead to desired impact performance [8, 17]. Within impact, frictional sliding along layer and fiber (inter)faces plays a role in energy absorption and might be hindered by a high level of adhesion. When a surface treatment is tailored for a macroscopic performance, it is advantageous that secondary properties remain unchanged, while the primary properties are improved [18]. Moreover, the performance under effects of harsh operation environments should remain acceptable [18, 19].

It was mentioned that high enough adhesion can lead to shattering of the aramid surfaces upon loading due to the internal structure of aramid fibers. Kanerva et al. [18] applied a diamond-like carbon (DLC) coating to form a nanoscale protective surface structure and also to gain high adhesion between the fibers and an epoxy matrix. The high adhesion related to DLC-coated aramid fibers and matrix polymers was also established by Devlin et al. (US Patent 6432537) [20] for short fibers. The latest research of modifying aramid fibers has been targeted to improve the internal structure of aramid fibers in order to prevent the fiber's cohesive damage at high interfacial loads [4].

4. Aramid fiber-matrix interfaces and tests

To understand the influence and nature of a modification on adhesion, the interface between fibers and matrix must be probed. Because most of the interest in surface modifications is founded on the targets by mechanical performance, the mechanical testing of the fiber-matrix bond is surveyed in the following.

Whenever a single filament or a bundle of fibers is analyzed, the samples or test specimens are small. Consequently, the load introduction and sensor configuration must be arranged in a highly sophisticated way. Currently, there are several test methods to study fiber-matrix interfaces:

- the micro-droplet method, e.g., [21];
- the single-fiber fragmentation method, e.g., [22];
- the pull-out method, e.g., [23];
- the fiber push-out method, e.g., [24]; and
- the fiber bundle pull-out method, e.g., [10].

The mechanical testing of fiber-matrix interfaces is not standardized, and therefore, the methodology among the current literature is not in harmony. The devices, specimen preparation, statistical significance, and the theory of data analysis vary in different reports and publications.

Due to the challenges of surfaces of aramid fibers, the tests of fiber-matrix interfaces are much used. The main presumption behind different test methods of interfaces is that the breakage of the interfacial bond occurs in a brittle manner. Also, it is typical to estimate that only shear load is subjected to the interface during testing. Then, the basic form of "interfacial strength" is denoted by interfacial shear strength (IFSS):

$$IFSS = \frac{F_{crit}}{A_{emb}} \tag{1}$$

where F_{crit} represents the peak value of the shear force observed during a test, and A_{emb} is the area carrying the load, i.e., the embedded fiber area. Eq. (1) is rather useful when the microdroplet method is used because the embedded area is relatively easy to determine. However, several corrections to the calculation of IFSS have been formulated when testing droplets bonded on carbon, aramid, glass, and natural fibers. Synthetic fibers, especially aramid fibers, are typically considered smooth or nearly smooth with only minor roughness that could result in sheer mechanical interlocking. Thus, the presumption of brittle failure of the interface ought to be justified.

The main deficiency related to interface tests is the lack of statistical significance and, consequently, reliability. For example, many of the works done using the microdroplet method are covering a low number of fiber-droplet samples and a low amount of variation in the droplet configuration per aramid fiber sample [25]. Also, the localized plastic deformation occurring in the droplet has an effect on the interfacial loading but is seldomly accounted for yet it can be detected easily in the microscopy images of the tested droplets, as shown in **Figure 1**.

A specific note related to aramid fibers is the role of friction between individual fibers and bundles in a fabric or preform structure. For certain ballistic applications, the amount of resin in the final product is low or entirely omitted, and then, the behavior of the aramid fiber-based reinforcement is governed by friction [26].

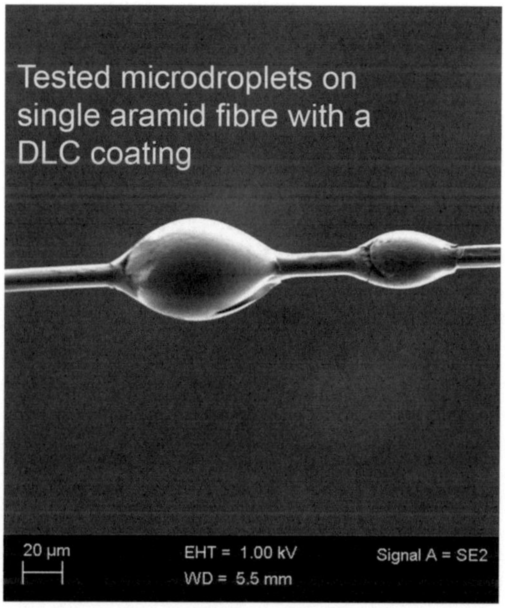

Figure 1.
An example of fiber-matrix droplets that have been tested using the microdroplet method [18].

In the end, the macroscopic behavior of a fibrous composite system is important. This means that the role of the interface and its strength should be known on the laminate level of length scales. Unfortunately, the exact relationship extending up from the single-filament behavior and up to the homogenized laminate level is lacking in the scientific literature. Several models exist to input interfacial effects while modeling bulk elasticity or strength [27]. To envisage the way that the combination of aramid fibers and matrix works, the well-known Halpin-Tsai model can be used as an example. For fibers, the model distinguishes between the effects of a particulate ("Greek symbol capital Phi" volume fraction) in its longitudinal direction and the effects in the transverse direction; these two directions are noted, by subindices L and T, here. Then, for the longitudinal direction, the model reads:

$$\frac{E_L}{E_m} = \frac{1 + \frac{l}{r}\eta_L \varphi}{1 - \eta_L \varphi} \tag{2}$$

where

$$\eta_L = \frac{\frac{E_f}{E_m} - 1}{\frac{E_f}{E_m} + \frac{l}{r}} \tag{3}$$

and where l is the fibrous particles' length, r is the diameter, and E refers to Young's modulus of the individual components. The individual components are the matrix (m) polymer and the fibers (f). For the transverse direction, the model reads:

$$\frac{E_T}{E_m} = \frac{1 + 2\eta_T \varphi}{1 - \eta_T \varphi} \tag{4}$$

where

$$\eta_T = \frac{\frac{E_f}{E_m} - 1}{\frac{E_f}{E_m} + 2} \tag{5}$$

By combining the longitudinal and transverse effects, the model yields the composite's stiffness (modulus) in a system of randomly oriented fibers:

$$E_c = a\,E_L + (1 - a)\,E_T \tag{6}$$

where a is a parameter that could be considered a function of the interface, orientation, or a shape factor. The a-parameter can be a constant value or a function of other external factors [28]. However, there is no universal model to implement interfacial effects specific to aramid fibers by the parameter.

For advanced composites, where the fibers are continuous, the above format of Halpin-Tsai equations cannot be used (l goes toward infinity in Eqs. 2 and 3). In this case, the bulk composite properties are entirely anisotropic, and the elastic constants, for example, must be determined for each of the three directions individually [2]. In these formulations, the so-called Halpin-Tsai parameters can be thought to represent the interfacial effects on the laminate's transverse and shear properties. Various studies have applied these equations to account for multiscale interfacial effects in composites (see, e.g., [29]).

5. Numerical predictions and finite element modeling

The experimental analysis of advanced composites is lacking the length scale of a representative bundle level. This is probably due to the practical challenges by small scale and due to the large variation, in size, of a representative bundle or bundle-bundle cross-over point. However, the numerical simulations can be harnessed with models on different length scales. Models of a single-length scale as well as multiscale routines are excellent tools to survey various effects on composite properties.

To model an interface, its volume in finite element (FE) models is commonly estimated to be zero, i.e., interface is a two-dimensional object or contact formulation. As a first estimate, this type of an interface can be estimated to behave in a brittle manner for aramid fibers, so that linear elastic fracture mechanics (LEFMs) are applicable in addition to sheer stress analysis. The power in the LEFM for interfaces is that a fracture toughness (G_i) in terms of a strain energy release rate can be used to describe the "strength" of an interface. With this type of a fracture parameter, the simulation results are somewhat less element mesh-dependent.

As noted, the plastic deformation of matrix around fibers affects interfacial breakage. To allow research of these effects, the models must be analyzed beyond LEFM. FE analysis with a homogenized interface model has been applied for models with a single fiber [30–32].

When plastic deformation at the interfacial region is considered, the fracture energy over the fiber surface (A_{emb}) divides into two parts upon fracture:

$$G_p = \frac{dW_p}{dA_{emb}} \tag{7}$$

and

$$G_e = \frac{dW_e}{dA_{emb}} \tag{8}$$

where the subindex p refers to plastic energy dissipation at interface, and the subindex e refers to the elastic strain energy release rate (ERR). In detail, the fracture toughness values can be related to the critical levels of energy release rate of damage onset (e.g., G_c) or propagation. In the applications of plastic dissipation at an interface, the common interface modeling method is the cohesive zone model (CZM). CZM refers to a mechanical model where the traction (τ) at the interface is defined as a function of the separation (δ) between the originally bonded bodies, i.e., fiber and matrix. As an example, a bilinear traction-separation law can be formulated as follows:

$$\tau = \begin{cases} K\delta & \delta \le a_0 \\ \frac{a_1 - \delta}{a_1 - a_0}\tau_0 \; for & a_0 \le \delta \le a_1 \\ 0 & \delta \ge a_1 \end{cases} \tag{9}$$

In Eq. (9), the below notations are used:

$$a_0 = \frac{\tau_0}{K} \tag{10}$$

and

$$a_1 = \frac{2G_c}{\tau_0}. \tag{11}$$

It can be seen that the simple bilinear formulation leads to two strength-related parameters: fracture toughness (G_c) and the critical traction (τ_0) related to the onset criterion of damage. Additionally, the numerical computation requires a definite value for the interfacial stiffness (K). Due to the strong relation to the computational procedures, the definition of K is vague from the point of view of material interfaces—several theories for CZM applications have been presented (see, e.g., [33]).

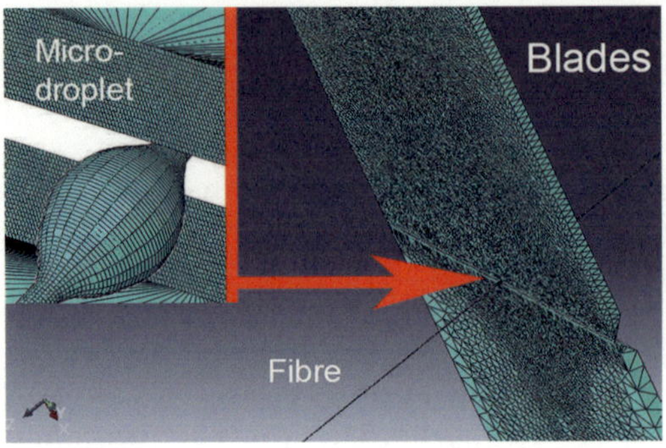

Figure 2.
An example of a 3D model simulating the microdroplet testing.

The power of numerical procedures allows to expand the analysis and consider exact three-dimensional models. Evidently, the parameters of the interface will have to defined in the three-dimensional system. For example, the damage onset will require a fracture criterion, and the fracture toughness will have to be applied via an interaction function.

For aramid fiber composites, an FE analysis was reported by Kanerva et al. [18] with a full 3D representation (see **Figure 2**). For the DLC-coated aramid fibers, values of 22.2 MPa and 500 J/m^2 were determined for the case-specific critical traction and interfacial fracture toughness, respectively.

It is clear that the current numerical modeling techniques and computational capacities can offer efficient tools to study fiber-matrix interactions in composites with aramid fibers. However, the multiplicity of parameters currently leads to overlapping fitting procedures. Thus, the solutions are not typically unitary to the simulation case in question. There is an urgent need to improve the microtest methods in order to gain more experimental output and data to validate the numerical models. There are very recent works in the current literature that target to improvements in the test systems to enhance statistical significance, data rate, and accurate output from the microtests [34, 35].

In future, it will be possible to accurately account for plastic deformation, residual stresses, and a multistage fracture process in the simulations of fiber-matrix interfaces. This will be an important step toward analyses of fatigue and environmental effects on interfaces in composites with aramid fibers.

6. Conclusions

Aramid fibers have been modified and optimized for various applications ever since they were established. Around three decades, academic researchers have focused on trial-and-error type experiments to improve the surface characteristics in the sense that feedback based on laminate testing was behind burdensome and expensive testing efforts. After millennia, the microlength scale testing of fiber-matrix interfaces has improved, and more information about the interfacial performance is available. Meanwhile, the numerical analysis of interfaces has gained increasing amounts of attention. Efficient and valid numerical simulations could save time and resources in the optimization of surface treatments of aramid fibers for specific applications. However, there are still challenges in the validation of numerical models because the experimental output does not provide for necessary parameters and statistical certainty.

Conflict of interest

The author confirms no conflicts of interest.

Author details

Mikko Kanerva
Materials Science and Environmental Engineering, Tampere University, Tampere, Finland

*Address all correspondence to: mikko.kanerva@tuni.fi

References

[1] Luon S, Van Ooij WJ. Surface modification of textile fibres for improvement of adhesion to polymeric matrices: A review. Journal of Adhesion Science and Technology. 2012;**16**:1715-1735

[2] Hull D, Clyne T. Elastic deformation of long-fibre composites. In: An Introduction to Composite Materials. Cambridge, United Kingdom: Cambridge University Press; 1996

[3] Roenbeck MR, Sandoz-Rosado E, Cline J, et al. Probing the internal structures of Kevlar fibers and their impacts on mechanical performance. Polymer. 2017;**128**:200-210

[4] Cheng Z, Yin Q, Wu H, He T, Luo L, Liu X. Regulating Cu(II)-benzimidazole coordination structure in rigid-rod aramid fiber and its composites enhancement effects. Composites Science and Technology. 2019;**184**:107837

[5] Garbassi F, Morra M, Occhiello E. Polymer Surfaces. NY: Wiley & Sons; 1994

[6] Maitland GC, Rigby M, Smith EB, Wakeman WA. Intermolecular Forces. Oxford: Clarendon Press; 1981

[7] Brown JR, Mathys Z. Plasma surface modification of advanced organic fibres. Journal of Materials Science. 1997;**3**:2599-2604

[8] Shaker M, Kamel I, Ko F, Song J. Improvement of the interfacial adhesion between Kevlar fiber and resin by using R-F plasma. Journal of Composites Technology and Research. 1996;**18**(4):249-255

[9] Lee-Sulivan P, Chian KS, Yue CY, Looi HC. Effects of bromination and hydrolysis treatments on the morphology and tensile properties of Kevlar-29 fibres. Journal of Materials Science Letters. 1994;**13**:305-309

[10] Palola S, Sarlin E, Kolahgar Azari S, Koutsos V, Vuorinen J. Microwave induced hierarchical nanostructures on aramid fibers and their influence on adhesion properties in a rubber matrix. Applied Surface Science. 2017;**410**:145-153

[11] Wu GM, Hung CH, Lu JC. Effects of plasma treatment on high performance fibres for composites. Sampe. 1999;**44**:1090-1097

[12] Lin J-S. Effect of surface modification by bromination and metalation on Kevlar fibre-epoxy adhesion. European Polymer Journal. 2002;**38**:79-86

[13] de Lange PJ, Akker P, Maas AJH, Knoester A, Brongersma HH. Adhesion activation of Twaron aramid fibres studied with low-energy ion scattering and x-ray photoelectron spectroscopy. Surface and Interface Analysis. 2001;**31**:1079-1084

[14] Tarantili PA, Andreopoulos AG. Mechanical properties of epoxies reinforced with chloride-treated aramid fibers. Journal of Applied Polymer Science. 1997;**65**:267-276

[15] de Lange PJ, Akker P, Mäder E, Gao SL, Prasithphol W, Young RJ. Controlled interfacial adhesion of Twaron aramid fibres in composites by the finish formulation. Composites Science and Technology. 2007;**67**:2027-2035

[16] Lin TK, Wu SJ, Lai JG, Shyu SS. The effect of chemical treatment on reinforcement/matrix interaction in Kevlar-fiber/bismaleimide composites. Composites Science and Technology. 2000;**60**:1873-1878

[17] Park SJ, Seo M-K, Ma T-J, Lee D-R. Effect of chemical treatment of Kevlar fibers on mechanical interfacial properties of composites. Journal of Colloid and Interface Science. 2002;**252**:249-255

[18] Kanerva M, Korkiakoski S, Lahtonen K, Jokinen J, Sarlin E, Palola S, et al. DLC-treated aramid-fibre composites: Tailoring nanoscale-coating for macroscale performance. Composites Science and Technology. 2019;**171**:62-69

[19] Imielińska K, Guillaumat L. The effect of water immersion ageing on low-velocity impact behaviour of woven aramid-glass fibre/epoxy composites. Composites Science and Technology. 2004;**64**:2271-2278

[20] Devlin D, Coates D, Archuleta T. Diamond-like-carbon coated aramid fibers having improved mechanical properties. US Patent 6432537; 2002

[21] Miller B, Baur U, Hirst DE. Measurement and mechanical aspects of the microbond pullout technique for obtaining fibre/resin interfacial shear strength. Composites Science and Technology. 1991;**42**:207-219

[22] Rao V, Herrera-Franco P, Ozzello AD, Drzal LT. A direct comparison of the fragmentation test and the microbond pull-out test for determining the interfacial shear strength. The Journal of Adhesion. 1991;**34**:65-77

[23] Favre JP, Merrine M-C. Characterisation of fibre/resin bonding in composites using a pull-out test. International Journal of Adhesion and Adhesives. 1981;**1**:311-316

[24] Mandell JF, Chen JH, McGarry FJ. A microdebonding test for in situ assessment of fibre/matrix bond strength in composite materials. International Journal of Adhesion and Adhesives. 1980;**1**:40-44

[25] Cen H, Kang Y, Lei Z, Qin Q, Qiu W. Micromechanics analysis of Kevlar-29 aramid fiber and epoxy resin microdroplet composite by micro-Raman spectroscopy. Composite Structures. 2006;**75**:532-538

[26] Sockalingam S, Chowdhury SC, Gillespie JW Jr, Keefe M. Recent advances in modeling and experiments of Kevlar ballistic fibrils, fibers, yarns and flexible woven textile fabrics – A review. Textile Research Journal. 2017;**87**(8):984-1010

[27] Guth E. Theory of filler reinforcement. Journal of Applied Physics. 1945;**16**:20-25

[28] Shokrieh MM, Moshrefzadeh-Sani H. On the constant parameters of Halpin-Tsai equation. Polymer. 2016;**106**:14-20

[29] Kanerva M, Jokinen J, Sarlin E, Pärnänen T, Lindgren M, Järventausta M, et al. Lower stiffness of GFRP after sulfuric acid-solution aging is due to degradation of fibre-matrix interfaces? Composite Structures. 2019;**212**:524-534

[30] Hodzic A, Kalyanasundaram S, Lowe A, Stachurski Z. The microdroplet test: Experimental and finite element analysis of the dependance of failure mode on droplet shape. Composite Interfaces. 1998;**6**(4):375-389

[31] Nishikawa M, Okabe T, Hemmi K, Takeda N. Micromechanical modeling of the microbond test to quantify the interfacial properties of fiber-reinforced composites. International Journal of Solids and Structures. 2008;**45**(14):4098-4113

[32] Sato M, Imai E, Koyanagi J, Ishida Y, Ogasawara T. Evaluation of the interfacial strength of carbon-fiber-reinforced temperature-resistant polymer composites by the micro-droplet test. Advanced Composite Materials. 2017;**26**(5):465-476

[33] Turon A, Dávila CG, Camanho PP, Costa J. An engineering solution for mesh size effects in the simulation of delamination using cohesive zone models. Engineering Fracture Mechanics. 2007;**74**:1665-1682

[34] von Essen M, Sarlin E, Tanhuanpää O, Kakkonen M, Laurikainen P, Hoikkanen M, et al. Automated high-throughput microbond tester for interfacial shear strength studies. In: Proceedings of the Sampe European Conference; 14-16 November 2017; Stuttgart, Germany: Sampe Europe; 2017

[35] Dsouza R, Jokinen J, Sarlin E, Antunes P, Kanerva M. Future microbond testing – finite element simulation of optical fibers for strains. In: Proceedings of the International Conference on Composite Materials (ICCM22); 11-16 August 2019; Melbourne, Australia. ICCM; 2019

Chapter 4

Vapor-Grown Carbon Fiber Synthesis, Properties, and Applications

J. Manivannan, S. Kalaiselvan and R. Padmavathi

Abstract

Carbon nanofibers are promising to revolutionize several fields in material science and are suggested to open the way into nanotechnology. Carbon fiber has become an important reinforcement material in composite materials and battery technology because of its low density, high strength, and tensile modulus. Furthermore, high electrical conductivity, thermal conductivity, and mechanical properties of carbon fiber make it useful in a wide variety of products. This chapter highlighting the synthesis and growth pattern of vapor grown carbon fiber (VGCFs). This chapter work reported here includes the application of VGCFs in composite and battery technology.

Keywords: VGCFs, composites, PAN, epoxy, polycarbonate, floating reactor

1. Introduction

Carbon nanofibers are linear filaments with diameter of 100 nm that are characterized by flexibility and their aspect ratio above 100. Materials as fiber are of extraordinary viable and scientific significance. The blend of high specific area, flexibility and high mechanical strength permit nanofibers to be utilized in our everyday life in addition creating arduous composites for vehicles and aviation. Be that as it may, they ought to be discriminated from conventional carbon fibers [1–3] in their small diameter (**Figure 1**). Conventional carbon fibers have a few micrometer-sized diameters.

Likewise, they are not quite the same as notable carbon nanotubes [5–9]. Carbon nanofibers could be developed by passing carbon feedstock nanosize estimated metal particles at high temperature [5, 10–15], which is fundamentally the same as the development state of carbon nanotubes.

Such a unique structure renders them to show semi-conducting behavior [16] and to have chemically active end planes on both the inner and outer surfaces of the nanofibers, thereby making them useful as supporting materials for catalysts [17], reinforcing fillers in polymeric composites [18], hybrid type filler in carbon fiber reinforced plastics [19–21], and photocurrent generators in photochemical cells [22, 23].

Though, their geometry is not quite the same as concentric carbon nanotubes containing a whole empty core, since they can be pictured as normally stacked shortened conical or planar layers along the filament length [24–27]. Such a one of a kind

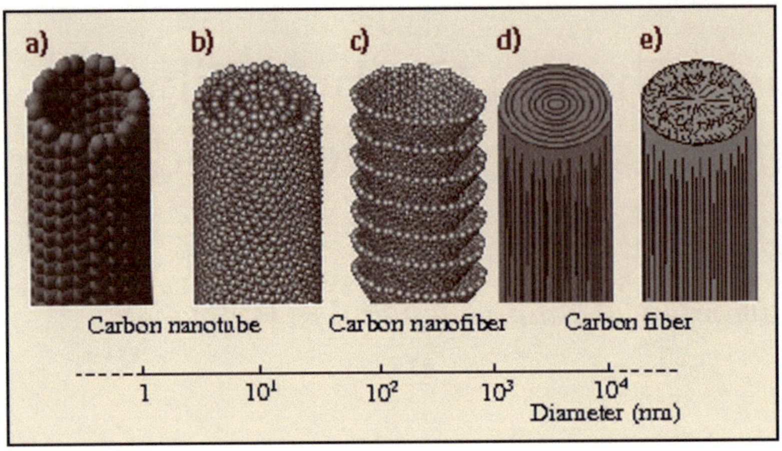

Figure 1.
Schematic comparison of the diameter dimensions for various types of fibrous carbons [4].

structure renders them to show semi- conducting behavior [16] and to have synthetically dynamic end planes on both the inner and outer surfaces of the nanofibers, in this way making them helpful as supporting materials for catalysts [17], reinforcing fillers in polymeric composites [18], hybrid type filler in carbon fiber reinforced plastics [19–21], and photocurrent generators in photochemical cells [22, 23].

2. Synthesis and properties of carbon fibers

Since the temperature and pressure essential to synthesize a carbon fiber from the fluid stage is at the triple point (Temperature—4100 K, Pressure—123 kbar), it would be practically difficult to get ready carbon fibers from the liquefy under modern preparing conditions. As a result, carbon fibers are arranged from organic precursors. This preparation is commonly done in three stages, including stabilization of a precursor fiber in air (at 300°C), carbonization at 1100°C, and subsequent graphitization (> 2500°C). Fibers undergoing only the first two steps are commonly called carbon fibers, while fibers undergoing all three steps are called graphite fibers.

Carbon fibers are commonly utilized for their high strength, while graphite fibers are utilized for their high modulus. Graphitic whiskers were developed under conditions close to the triple point of graphite. At that point, the structural model was proposed, in which the layers comprising of graphene sheets are twisted around the axis like as in rolling up a carpet. These whiskers were utilized as the presentation focus in the beginning times of carbon fiber technology, despite the fact that they have never been manufactured on a large scale.

Carbon fibers are

1. High-strength polyacrylonitrile (PAN)-based fiber

2. A high-modulus PAN-based fiber

3. A mesophase pitch-based carbon fiber (MPCF).

The PAN-based fibers comprise of little sp^2-carbon structural units specially lined up with the carbon hexagonal portions corresponding to the fiber axis. This orientation is liable for the tensile strength of PAN-based carbon fibers [28].

By shifting the preparing conditions (e.g., oxidation conditions, choosing of precursor material, and particularly by increasing the heat treatment temperature) of PAN fibers, a better arrangement of the graphene layers can be accomplished, thus leading to stiffer, higher-modulus PAN fibers, however with lower strength [29]. PAN-based fibers are one of the distinctive hard carbons.

MPCFs comprise of perfectly-aligned graphitic layers equivalent to the corresponding fiber axis, and this high level of favored direction is liable for their high modulus or stiffness just as their high graphitizability. The structures depicted above suggest ascent to various physical properties, although each kind of fiber highlights carbon hexagonal systems, having the sturdiest covalent bonds in nature (C–C bonds). These sturdy interatomic bonds lie in sheets basically corresponding to the fiber axis, and are liable for the high mechanical performance of these carbon fibers.

3. Vapor-grown carbon fibers

PAN- based fibers have high strength and MPCFs have high modulus, while VGCFs afford fundamentally ultra-high modulus materials. VGCFs have an extremely unique structure like annular-rings and are prepared by something diverse arrangement process than that used to prepare PAN-based and MPCFs. Specifically, VGCFs are not prepared from a fibrous precursor, but instead from hydrocarbon gas, utilizing a catalytic growth process [5, 11–15]. Ultrafine transition metal particles, for example, iron particles with diameter under 10 nm, are scattered on a ceramic substrate, and a hydrocarbon, for example, benzene diluted with hydrogen gas is introduced at temperatures of around 1100°C. Hydrocarbon decomposition happens on the catalytic particle, showing a ceaseless carbon take-up by the catalytic particle and a persistent output by the particle of efficient tubular filaments of hexagonal sp^2 carbon. The swift development rate which is 106 times quicker than that analyzed for the development of normal metal whiskers [30], permits the production of commercially suitable amounts of VGCFs.

4. Vapor grown carbon Fiber growth

While portraying the production of VGCF, two distinct strategies can be discriminated. In the primary strategy, the fibers are manufactured on catalyst-seeded substrates, in two independent consecutive stages. In the first stage, the impetus molecule, for the most part with iron being the prevalent constituent, starts a long, thin, somewhat graphitic fiber when presented to a hydrocarbon gas close or more 1000°C.

A petite fraction of these fibers grows to macroscopic lengths when exhibited to a low carburizing potential gas, while keeping up the outside diameter of the initial catalytic particle. The sizes of these particles, and subsequently the filament diameter, have been considered from 10 to 200 nm [31–33]. The development of the filament may extend as quickly as 1 mm/min and persevering for a few minutes until the catalytic particle is deactivated [33]. At this stage filament lengths of a few centimeters might be acquired.

In the second stage, when the gas potential is increased, the filament thickens because of the deposition of pyrolytic carbon. Fibers with diameters up to 100 nm have been produced in batch conditions by this method. Pyrolytic carbon is deposited with the basal planes specially arranged corresponding to the surface, the properties of the fiber are moderately graphitic [34].

Methane, hydrogen and an inert gas are utilized to develop the fibers at atmospheric pressures on nesting cylindrical substrates inside a growth tube. Lower centralizations of the methane-hydrogen blends are utilized for fiber extending, though higher concentrations are utilized for thickening. As methane is an economical feedstock in certain parts of the world, the production of VGCFs can turn out to be exceptionally financial contrasted with other carbon filaments.

5. Mass production of vapor-grown carbon nanofibers (VGCFs)

On account of their unique properties, for example, high strength and electric conductivity and special functional properties, researchers have indicated a lot of consideration for the large scale manufacturing of these materials.

1. The physical properties of vapor grown carbon fibers, with littler diameters across ought to be more grounded than those with larger diameters fume developed carbon filaments and exploratory outcomes demonstrated this expectation.

2. Furthermore, it was accounted for that the distance across of the nanofibers is represented by the size of the impetus particles.

3. For the large scale manufacturing of VGCFs the key procedure is the seeding of the impetus particles. VGCFs with distances across of 50–100 nm was created effectively utilizing a blend of ferrocene and fluid hydrocarbon in a vertical heater. The outcomes show that this strategy gives the chance of large scale manufacturing. Be that as it may, in view of the trouble of proficiently scattering little Fe particles in showering impetus arrangement, vague blends of filaments and residue were constantly gotten.

Carbon nanofibers were manufactured with diameters of 50–80 nm from catalytic decomposition of hydrocarbon gas with catalyst metal powder utilizing the support catalyst method. In spite of the fact that the support catalyst technique can create high quality fibers, the preparation and defined scattering of ultrafine catalyst particles are very difficult and are exceptionally troublesome, the yield of this strategy is spendthrift to be in any way marketed. So as to tackle the above issues the floating catalyst technique was created with the help of support catalyst to deliver carbon nanofibers with diameter of 50–100 nm.

6. Vapor grown carbon nanofiber by floating reactor method

Vapor-grown carbon fibers (VGCFs) have been developed by the deterioration of hydrocarbons, for example, benzene and methane, utilizing transition metal particles as a catalyst at a development temperature of 1000–1300°C [31–35]. These fibers have been portrayed regarding the profoundly favored orientation of their graphitic basal planes corresponding to the fiber pivot, with an annular ring surface in the cross segment. This structure offers ascend to amazing mechanical properties, high electrical and thermal conductivity, and a high graphitizability of the fibers [32, 33]. In this way, numerous scientists have attempted to arrive at the last objective of large scale manufacturing of these fibers filaments with ease.

Among the different preparing techniques, the most significant one has been the advancement of the floating reactant technique [36, 37], which permits a three-dimensional scattering of the hydrocarbon together with the catalytic particles got from the pyrolysis of organometallic compounds, for example, ferrocene, in a reaction chamber, ensuing about a high yield and a fairly uniform diameter across of the resulting fibers. Thus, the floating reactant strategy is believed to be a promising method for the large scale manufacturing of carbon fibers at generally economical.

7. Fiber morphology

Because of their production method, VGCFs arrive in an assortment of morphologies that are fundamental to understand their properties. A few unique morphologies have been accounted in the literature [38]. Somewhat, these morphologies emulate the internal structure of the fiber. The structure of VGCFs looks like that of a tree trunk, with concentric annular rings. The original filament lies in the middle of axis of symmetry. The layers are in part aligned turbostratic carbon, with the average basal plane skewed from the fiber axis by around 10–15° [30]. In thinner fibers, or in fibers submitted to high temperature heat-treatments, cracks are observed between consecutive rings, and the external surfaces are often crenulated. Cracks are seen between consecutive rings when thinner fibers, or in fibers exposed to high temperature, and the external surfaces are frequently crenulated.

When the pyrolytic carbon is subjected to cooling, it shrinks more quickly in the radial than in the tangential direction. The ensuing radial stress (tension) will be reduced if circumferential cracking between the rings happens. Then again, the contraction in the tangential direction prompts a high compressive pressure that can be halfway eased by the formation of the crenulations.

The uniform geometry and the basically carbon structure of the VGCFs makes the fibers very graphitizable. Swift graphitization increased with increased temperatures [28]. Above 2600°C, VGCFs structure moves toward like a single crystal graphite structure. Evidently, this property will let improvement to a specific level, tailoring of the fiber properties to preferred values. In recent years, a widen interest has been appeared in the surface properties of VGCFs, as this can be identified with the interfacial properties of the fiber in composites. It was pointed out that the surface reactivity of VGCFs is lower than that of ex-pitch or ex-PAN carbon fibers [39]. On the other hand, the surface reactivity can be increased by oxidation treatment to be nearer to that of the other carbon fibers.

The effect of diverse oxidative surface treatments (nitric acid, plasma, air and carbon dioxide) on the fibers surface reactivity has been studied [30].

It was observed that, surface oxidation of carbon fibers is the standard innovation to increase adhesion between fiber and matrix, air and carbon dioxide treatments do not fundamentally expand the fiber's surface reactivity, in spite of extensive weight reduction. This was ascribed to the presence of traces of iron left on the VGCF from the growth stages, that may have catalyzed carbon gasification by a pitting mechanism [40].

Cleansing of the VGCFs with HCl to remove the iron can prevent this behavior. Although nitric acid and plasma treatments were found to increase surface reactivity without altering significantly the morphology of the fibers, the latter appeared to be more favorable in improving the adhesion of VGCFs to a matrix.

To expel the iron, VGCFs is cleaned with HCl can obviate this behavior. Though nitric acid and plasma treatments were found to increase surface reactivity without changing fundamentally the morphology of the fibers, the latter gave off an impression of being progressively great in improving the adhesion of VGCFs to a network.

8. Properties of VGCFs

8.1 Mechanical properties

The modulus and the elasticity of VGCFs can be found easily [34]. The calculated modulus value varies from 100 to over 1000 GPa. The stiffness of VGCF depends upon their diameters. Even though the fibers are produced in the same experiment the diameters increased from 6 to 32 nm, the modulus decreases 300–1200GPa [33, 34].

The degree of favored orientation of graphitic basal planes is related to the stiffness of pyrolytic carbon. The stiffness of a graphite fiber is connected as a component of the orientation point of the graphitic plane. This model was fitted to the stiffness-data of VGCFs by measuring their orientation parameter. This model was fitted to the solidness information of VGCFs by estimating their direction parameter. As the main fitting parameter utilized in the model is free of fiber diameter, the dwindle in the solidness of thicker fibers was assigned to the decrease in their graphitic requesting.

This again is because of a swifter deposition of pyrolytic carbon during the thickening period. This is as per the finding that that increasing the graphitization of VGCF by heat-treatment also additionally increases their modulus considerably. In one example, the modulus of vapor grown fibers more than doubled to about 500 GPa when they were heated at 2200°C [30].

Heat-treated fibers are unsuccessful more commonly in the "sword-in-sheath" mode and in this mode; grouping of circumferential breaks along the internal rings permits continuous cylinders to slide telescopically inside one another, slowly diminishing the load bearing limit of the fiber. It is obvious from these outcomes that the modulus and the disappointment method of vapor grown carbon fibers are connected to their structure.

The values of the tensile strength of VGCF are progressively reliable, changing from 2.5 to 3.5 GPa for fibers with diameters across just beneath 10 μm. The reliance on the diameter however is strong [33, 34, 41]. This was clarified by accepting that thicker fibers have a larger flaws population, and a more prominent likelihood of failure, than thinner ones.

8.2 Electrical properties

The carbon structure and potential graphitization of VGCFs by heat-treatment makes them a fascinating possibility for thermal and electrical applications. VGCFs are known to have the most noteworthy electrical and thermal conductivities among carbon fibers because of their conceivable high structural perfection [42].

It was verified that VGCFs heat-treated at lower temperature have a resistivity that is practically constant with temperature. The resistivity of fibers heat-treated at higher temperatures decreases with the operating temperature. At the most elevated temperatures, the resistivity of the VGCFs approaches that of single crystal graphite, with a resistivity of about 5×10^{-7} ohm cm at 300 K. Besides, the decrease in resistivity with fiber diameter across is ascribed to the expanded arranged graphite regions as the diameter increases.

8.3 Thermal properties

The thermal conductivity of VGCFs is astoundingly high [31]. The thermal conductivity of both s-grown VGCFs and the other VGCFs heat-treated to 3000°C were evaluated through a temperature range of 10–300 K. It was demonstrated that the heat treatment expands the fiber conductivity by a factor of 50. The heat-treated fibers rank among the best thermal conductors available, at room temperature.

Besides, it was demonstrated an immediate connection between electrical resistivity, thermal conductivity and tensile modulus of carbon fibers. The strong correlation between thermal conductivity and electrical resistivity was elucidated by comparative fiber structural effects dominating these properties. This, in blend with the recently pointed out connection between fiber structure and tensile modulus [34], could prompt the expectation of the properties legitimately from fiber structural considerations.

9. Applications of VGCFs

9.1 In composites

In recent times, there has been a developing enthusiasm for the utilizations of VGCF in carbon and polymer matrix composites. The primary thermoplastic submicron-size VGCF composites, utilizing polycarbonate and nylon as a matrix was reported [36]. The growing mechanical and thermal properties were calculated with fiber volume fraction, despite the fact that this increase is less than could be theoretically expected. The composites were compression molded utilizing a straightforward set-up bringing about a fiber arrangement that was believed to be irregular in three dimensions. It was anticipated higher properties if a process like injection molding was utilized to stimulate fiber alignment.

The process ability of submicron VGCFs in thermoplastic matrices was analyzed by extruding polycarbonate-VGCF composites in a single screw extruder and subsequently melt-drew the composites to induce fiber orientation. It was concluded that the dispersion of the fibers in the matrix was very poor at 2 vol% fiber fraction and turned out to be much less fortunate when the fiber content was expanded to 5 vol%. It was recommended that grinding or melt-blending the fiber with the polymer before extrusion may bring better result in better dispersion.

The utilization of submicron VGCFs as arbitrary layers between layers of ceaseless carbon fibers in an epoxy matrix was studied [43] and it was found that the VGCFs improve appreciably the damping capacity of the composites at fiber fractions as low as 0.6 vol%.

The utilization of VGCFs for fabricating thin paper was made effectively by supplanting the generally utilized commercial chopped ex-PAN and ex-pitch fibers to acquire planar isotropy [44].

Submicron VGCFs were used in cement-matrix composites, an application wherein they could be amazingly valuable, because of the relative minimum cost and simple process ability [41]. With a fiber heap of just 1.5 vol%, they accomplished the most noteworthy electromagnetic interference (EMI) shielding effectiveness ever achieved for a cement-matrix composite.

Composites based on VGCF-mats were made and heat-treated at 2800°C and their thermal conductivity was studied [45]. It was acquired that the 36 vol% composite exhibited a room-temperature thermal conductivity of 564 W/mK. This value is some extent higher than that of copper (+/− 450 W/mK) and it was ascribed to the greatly graphitic nature of the VGCF mats.

Aluminum matrix and carbon-carbon composites based on VGCFs were fabricated and compared [46]. For the aluminum-VGCF composites at 36.5 vol% the thermal conductivity of 642 W/mK was found. An outstanding value of 910 W/mK was observed for a 70 vol% carbon-VGCF composite which is double the value of copper.

S.No	Properties	Carbonized s-VGCFs[a]	Graphitized s-VGCFs[b]	Method needed for characterization
1	Lattice constant	6.900 A	6.775 A	XRD
2	Diameter of fiber	0.2 mm	0.2 mm	SEM
3	Length of fiber	10–20 mm	10–20 mm	SEM
4	Volume density	0.02–0.07 g/cm^3	0.02–0.07 g/cm^3	Tapping
5	Real density	1.9 g/cm^3	2.1 g/cm^3	Pycnometer
6	Surface area (BET)	37 m^2/g	15 m^2/g	N absorption
7	Ash content	1.5%	0.03%	SDK
8	pH	5	7	PH Meter
9	Starting temperature of oxidation	550°C	650°C	TGA

[a]Carbonized s-VGCFs indicate the sample heat treated at 1200°C.
[b]Graphitized s-VGCFs indicate the sample heat treated at 2800°C.

Table 1.
Basic properties of submicron VGCFs [48].

The mechanical properties of VGCF-carbon composites was studied and arrived the result that the tensile properties of these composites are lower than anticipated It was concluded that the thermal properties of VGCFs are preferred and converted into composites.

9.2 Application in batteries

The fibers acquired by the floating reactant method have a thin diameter distribution (from 0.1 to 0.2 mm), indicating nearly a similar morphology when contrasted with that of normal VGCFs (10–20 mm), which comprise of a central filament and an external deposit of pyrolytic carbon with the annular structure of a tree. Moreover, these fibers have a high possibility for application as filler in composites and as an anode material in lithium ion batteries, including additives to anode materials, because of incredible conductivity and high surface to volume ratio.

For s-VGCFs acquired by a floating reactant technique, the fundamental properties and micro structural development of these fibers with Heat Treatment Temperature (HTT), the physical properties of a solitary fiber and in the massive state were assessed to apply this material as the filler in electrodes of lead acid batteries and in the Li-ion battery system. Graphitized s-VGCFs showed attractive desirable properties when utilized as filler in electrodes, for example, genuinely high mechanical strength and electrical conductivity in a single fiber, a high degree of resiliency and good volume conductivity in the bulky state. These desirable properties should yield improved execution performance in actual batteries (**Table 1**) [47, 49–51].

10. Conclusion

Vapor grown carbon fiber composites for batteries and supercapacitors have been extensively studied worldwide. The CNFs and their composites can be utilized in numerous fields; they have unique mechanical, electrical and thermal properties that may be useful in making composite materials like polycarbonate-VGCF, epoxy-VGCF, PAN-VGCF and aluminum-VGCF and filler in electrodes of lead-acid batteries and the Li-ion battery system.

Author details

J. Manivannan[1], S. Kalaiselvan[1*] and R. Padmavathi[2]

1 Department of Chemistry, SNS College of Technology, Coimbatore, Tamilnadu, India

2 Department of Chemistry, M. Kumarasamy College of Engineering, Karur, Tamilnadu, India

*Address all correspondence to: kalaichem82@gmail.com

References

[1] Donnet JB, Bansal RC. Carbon Fibers (Marcel Dekker, New York). Journal of Polymer Science Part C: Polymer Letters. 1984;**23**(5)

[2] Peebles LH. Carbon Fibers. Boca Raton: CRC; 1994

[3] Chung DDL. Carbon Fiber Composites. Boston: Butterworth Heinemann; 1994

[4] Kim YA, Hayashi T, Endo M, Dresselhaus MS. Carbon Nanofibers. In: Vajtai R, editor. Springer Handbook of Nanomaterials. Berlin, Heidelberg: Springer Handbooks. Springer; 2013

[5] Oberlin A, Endo M, Koyama T. Filamentous growth of carbon through benzene decomposition. Journal of Crystal Growth. 1976;**32**:335-349

[6] Iijima S. Helical microtubules of graphitic carbon. Nature. 1991;**354**:56-58

[7] Dresselhaus MS, Dresselhaus G, Eklund P. Science of Fullerenes and Carbon Nanotubes. New York: Academic; 1996

[8] Saito R, Dresselhaus G, Dresselhaus DS. Physical Properties of Carbon Nanotubes. London: Imperial College Press; 1998

[9] Ajayan PM, Zhou OZ. Applications of carbon nanotubes. In: Dresselhaus MS, Dresselhaus G, Avouris P, editors. Carbon Nanotubes. Topics in Applied Physics. Vol. 80. Berlin, Heidelberg: Springer; 2001

[10] Dresselhaus MS, Dresselhaus G, Sugihara K, Spain IL, Goldberg HA. Graphite Fiber and Filaments I. Berlin Heidelberg: Springer; 1988

[11] Baker RTK. Catalytic growth of carbon filaments. Carbon. 1989;**27**:315-323

[12] Endo M. Grow carbon fibers in the vapor phase. Chemical Technology. 1988;**18**:568-576

[13] Tibbetts GG. Why are carbon filaments tubular? Journal of Crystal Growth. 1984;**66**:632-637

[14] Rodriguez NM, Chambers A, Baker RTK. Catalytic engineering of carbon nanostructures. Langmuir. 1995;**11**:3862-3866

[15] Tibbetts GG. Vapor-grown carbon fibers: Status and prospects. Carbon. 1989;**27**:745-747

[16] Endo M, Kim YA, Ezaka M, Osada K, Yanagisawa T, Hayashi T, et al. Selective and efficient impregnation of metal nanoparticles on cup-stacked-type nanofibers. Nano Letters. 2003;**3**:723-726

[17] Choi YK, Gotoh Y, Sugimoto KI, Song SM, Yanagisawa T, Endo M. Processing and characterization of epoxy nanocomposites reinforced by cup-stacked carbon nanotubes. Polymer. 2005;**46**:11489-11498

[18] Yokozeki T, Iwahori Y, Ishiwata S. Matrix cracking behaviors in carbon fiber/epoxy laminates filled with cup-stacked carbon nanotubes (CSCNTs). Composites Part A: Applied Science and Manufacturing. 2007;**38**:917-924

[19] Yokozeki T, Iwahori Y, Ishiwata S, Enomoto K. Mechanical properties of CFRP laminates manufactured from unidirectional prepregs using CSCNT-dispersed epoxy. Composites Part A: Applied Science and Manufacturing. 2007;**38**:2121-2130

[20] Yokozeki T, Iwahori Y, Ishibashi M, Yanagisawa T, Imai K, Arai M, et al. Fracture toughness improvement of CFRP laminates by dispersion of cup-stacked carbon nanotubes.

Composites Science and Technology. 2009;**69**:2268-2273

[21] Saito K, Ohtani M, Fukuzumi F. Electron-transfer reduction of cup-stacked carbon nanotubes affording cup-shaped carbons with controlled diameter and size. Journal of the American Chemical Society. 2006;**128**:14216-14217

[22] Norris ID, Shaker MM, Ko FK, MacDiarmid AG. Electrostatic fabrication of ultrafine conducting fibers: Polyaniline/polyethylene oxide blends. Synthetic Metals. 2000;**114**:109-114

[23] Bacon R. Production of graphite whiskers. Journal of Applied Physics. 1960;**31**:283-290

[24] Endo M, Kim YA, Fukai T, Hayashi T, Oshida K, Terrones M, et al. Structural characterization of cup-stacked type nanofibers with an entire hollow core. Applied Physics Letters. 2002;**80**:1267-1269

[25] Yoon SH, Lim S, Song Y, Ota Y, Qiao WM, Tanaka I. Mochida: KOH activation of carbon nanofibers. Carbon. 2004;**42**:1723-1729

[26] Yoon SH, Park CW, Yang HJ, Korai Y, Mochida I, Baker RTK, et al. Novel car-bon nanofibers of high graphitization as anodic materials for lithium ion secondary batteries. Carbon. 2004;**42**:21-32

[27] Liu QF, Ren WC, Cheng ZG. Semiconducting properties of cup-stacked carbon nanotubes. Carbon. 2009;**47**:731-736

[28] Endo M, Koyama T, Hishiyama YM. Structural improvement of carbon fibers prepared from benzene. Japanese Journal of Applied Physics. 1976;**15**:2073

[29] Serp P, Figueiredo JL, Bernardo CA. In: Palmer KR, Marx DT, Wright MA, editors. Carbon and Carbonaceous Composite Materials. Vol. 134. Singapore: World Scientific Publishing; 1996

[30] Serf P, Figueiredo JL. An investigation of vapor-grown carbon fiber behavior towards air oxidation. Carbon. 1997;**35**(5):675-683

[31] Koyama T, Endo M. Electrical resistivity of carbon fibers prepared from benzene. Japanese Journal of Applied Physics. 1974;**3**(7):1175-1176

[32] Speck JS, Endo M, Dresselhaus MS. Structure and intercalation of thin benzene derived carbon fibers. Journal of Crystal Growth. 1989;**94**:834-848

[33] Tibbetts GG. Carbon fibers produced by pyrolysis of natural gas in stainless steel tubes. Applied Physics Letters. 1983;**42**(8):666-668

[34] Oberlin A, Endo M. Filamentous growth of carbon through benzene composition. Journal of Crystal Growth. 1976;**32**:335-349

[35] Koyama T, Endo M. Structure and properties of graphitized carbon fiber. Japanese Journal of Applied Physics. 1974;**13**(12):1933-1939

[36] Benissad F, Gadelle P, Coulon M, Bonnetain L. Formation de fibers de carbone a partidu menthane: I croissane catalytique et epaississement pyrolytique. Carbon. 1988;**26**(1):61-69

[37] Ishioka M, Okada T, Matsubara K, Endo M. Formation of vapor grown carbon fibers mixtures. II. Influence of catalyst. Carbon. 1992;**30**(6):865-868

[38] Tibbetts GG, Beetz CP Jr. Mechanical properties of vapour-grown carbon fibres. Journal of Physics D: Applied Physics. 1987;**20**(3):292

[39] Darmstadt H, Roy C, Kaliaguine S, Ting J-M, Alig RL. Surface spectroscopic

analysis of vapour grown carbon fibres prepared under various conditions. Carbon. 1998;**36**(7-8):118-1190

[40] Koyama T, Endo M, Hishiyama Y. Structure and properties of graphitized carbon Fiber. Jap. J. Appl. Phys. 1974;**13**(12):1933

[41] Xuli F, Chung DDL. Submicron-diameter-carbon-filament-cement-matrix composites. Carbon. 1998;**36**(4):459-462

[42] Endo M, Komaki K. Extended abstracts of 16th Biennial Conference on Carbon. San Diego, CA: American Carbon Society; 1983. p. 523

[43] Hudnut SW, Chung DDL. Carbon. 1995;**33**:1627

[44] Guth JR, Hart DW, Ting J-M. Extended abstracts of 22nd Biennial Conference on Carbon. San Diego, CA: American Carbon Society; 1995. p. 288

[45] Ting J-M, Lake ML, Duffy DR. Journal of Materials Research. 1995;**10**:1478

[46] Ting J-M, Guth JR. Extended abstracts of 22nd Biennial Conference on Carbon. San Diego, CA: American Carbon Society; 1995. p. 296

[47] Endo M, Kim YA, Hayashi T, Nishimura K, Matusita T, Miyashita K, et al. Vapor-grown carbon fibers (VGCFs): Basic properties and their battery applications. Carbon. 2001;**39**:1287-1297

[48] Endo M, Kim YJ, Hayashi T, Nishimura K, Matusita T, Miyashita K, et al. Vapor-grown carbon fibers (VGCFs): Basic properties and their battery applications. Materials Science. 2001

[49] Endo M, Saito R, Dresselhaus MS, Dresselhaus G. From carbon fibers to carbon nanotubes. In: Ebbesen TW, editor. Carbon Nanotubes. New York: CRC; 1997. pp. 35-105

[50] Oberlin A. High-resolution TEM studies of carbonization and graphitization. Chemistry and Physics of Carbon. 1989;**22**:1-135

[51] Koyama T, Endo M. Ohyo Butsuri. 1973;**42**:690

Chapter 5

Fiber Composites Made of Low-Dimensional Carbon Materials

Yan Xu and Xian Zhang

Abstract

In recent years, a scientific shift has been observed that the use of carbon-based nanomaterials in different composite materials can improve their mechanical, thermal, and electrical properties. Different carbon-based nanomaterials have various structures and mechanical, electrical, and thermal conductivity characteristics. By combining with different material composite methods, carbon composite materials with different structures can be prepared. Through the optimization of material structure, carbon composite materials with high performance can be obtained. SP^2 hybrid carbon materials, such as carbon nanotubes (CNTs), carbon fiber (CF), and graphene, have excellent electrical, thermal, and mechanical properties due to their regular carbon six-membered ring structure, so they are the main low-dimensional carbon materials and are widely used in composite research. In this chapter, the research progress of carbon nanotubes, carbon fibers, and graphene-based fibers (GBFs) in composite materials are introduced, respectively, and the preparation method, molding process, performance, and application in industry are summarized. Finally, the existing problems and future development trend of carbon-based composites are prospected.

Keywords: carbon-based nanomaterials, carbon nanotubes (CNTs), carbon fibers (CF), graphene-based fibers (GBFs), fiber composites

1. Introduction

Among all kinds of low-dimensional materials [1–8], using carbon-based low-dimensional materials to improve their physical, mechanical, and electrical properties has become a trend [9]. These carbon-based nano/micron additives include carbon fibers (CF), single-walled and multi-walled carbon nanotubes (SWCNTs and MWCNTs) [10–12], graphene oxide (GO) [13–15], and graphene nanoplates (GNP) [16]. The results show that the composites with high strength ductility, dimensional stability, and economy can be produced. They have an attractive application prospect in the fields of microelectronic devices, aerospace, energy, chemical industry, etc.

Different carbon material units have various structures and mechanical, thermal, and electrical properties. By combining with different material composite methods, carbon composite materials with different structures can be prepared. Through the optimization of material structure, carbon composite materials with high performance can be obtained [17, 18]. This chapter mainly introduces the preparation methods, properties, and application fields of carbon-based nanomaterials, such as CNTs,

CF, and GBFs, which are commonly used to assemble macro carbon composites. The preparation methods of GBFs and their composite fibers, as well as their applications in sensors, energy storage, energy conversion, and other aspects, such as supercapacitors, lithium-ion batteries (LIBs), actuators, and solar cells, are mainly introduced. Finally, the existing problems and future development of carbon-matrix composites are summarized.

2. Carbon nanotubes (CNTs)

CNTs were first discovered under TEM in 1991. It is a one-dimensional tubular material made of SP^2 hybrid carbon atoms. Its diameter ranges from several nanometers to tens of nanometers, and its length can reach centimeter-level at most. According to the wall layer, it can be divided into single-walled CNTs (SWCNTs) and multi-walled CNTs (MWCNTs) (**Figure 1**). It is the most commercialized nanofiber with the highest strength and the smallest diameter [19–21]. Moreover, CNTs have good toughness, which can withstand 40% of tensile strain without brittle behavior or fracture phenomenon, thus improving the toughness of matrix composite [22]. CNTs with super high aspect ratio and excellent mechanical and physical properties, such as high strength, high thermal conductivity, high conductivity, and low thermal expansion coefficient, are regarded as the ideal functional modifier for preparing high-performance composite materials [23–25].

2.1 Fabrication of CNTs

The preparation methods of CNTs include chemical vapor deposition (CVD), arc discharge (AD), and laser ablation (LA) [26]. CVD is the most commonly used method to prepare CNTs in the laboratory. Generally, CNTs are grown under the action of the catalyst after carbon source cracking at a certain temperature. This method has a series of advantages, such as simple equipment, fast preparation speed, large output, and controllable quality. The catalysts are generally transition metals such as iron, cobalt, and nickel, and the carbon sources are generally carbon-containing organics such as methane, ethylene, acetylene, ethanol, and xylene. The morphology (diameter, wall layer, length, density, curvature, crystallinity, etc.) of CNTs can be tuned by controlling the type and concentration of catalyst, the ratio of carbon source and injection speed, the temperature, pressure, and time of CVD [27–30].

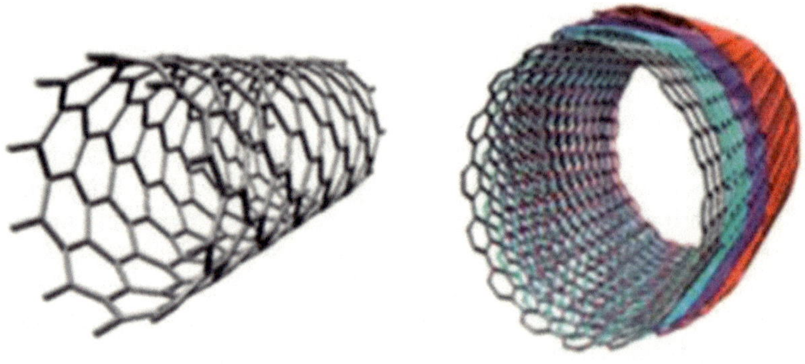

Figure 1.
Schematic diagrams of fullerene single-walled carbon nanotube (SWCNT) and multi-walled carbon nanotube (MWCNT) [19].

The AD method is also the main method to produce CNTs. Usually, in the low-pressure arc chamber of inert gas, hydrogen, or other gases, the graphite material is used as the electrode to generate a continuous arc between the electrodes, which makes the graphite react with the catalyst to generate CNTs. The AD method has a high yield, and the CNT's crystal structure is relatively complete [31, 32].

Laser ablation is a method to prepare CNTs by bombarding the surface of graphite target doped with iron, cobalt, nickel, and other transition metals in an inert gas environment at 1200°C [33]. The advantage of this method is that the CNTs produced are of high purity and convenient for continuous production, but this method is not suitable for large-scale macro production due to its high energy consumption, complex equipment, and high preparation cost [34, 35]. In addition to the above three main preparation methods, CNTs can also be prepared by template method, flame method, solar energy method, and electrolytic alkali metal halide method [36].

2.2 Properties of CNTs

Because of its special tubular structure and the strong binding force between SP^2 hybrid carbon atoms, CNTs have high strength, fracture toughness, and elastic modulus, which are superior to any one-dimensional fiber [37]. The tensile strength of CNTs can reach 50–800 GPa, nearly 100 times of that standard steel, about 200 times higher than that of other polymer fibers, and its structure can be kept intact under 1 million atmospheric pressure. CNTs will not break obviously under large bending, while graphite fiber will break when bending 1% (volume fraction). The maximum elastic modulus of CNTs is 1 TPa, which is equivalent to that of diamond and about five times to that of steel. Due to defects, the actual elastic modulus of MWCNTs is in the range of 20–50 GPa [38–40]. Fiber is usually used to strengthen composite materials. In addition to its own strength, a high aspect ratio (>20) is also a key factor to obtain high-strength composite materials. The aspect ratio of CNTs is generally >1000. Therefore, through CNT-reinforced composite materials, it can show good mechanical strength and fatigue resistance [10, 41].

The carbon atoms in CNTs are arranged in a six-membered ring network structure, which is very conducive to phonon vibration. Therefore, CNTs have good thermal conductivity. Due to the anisotropy of the structure, the thermal conductivity of CNTs along the length direction is much higher than that in the vertical direction. Theoretically, the thermal conductivity of SWCNTs can reach 10,000 W/mK at room temperature. Due to the presence of impurities, the highest experimental values of SWCNTs and MWCNTs are 3500 and 3000 W/mK, respectively [42–44]. Theoretical calculation and experimental results show that with the increase of CNT diameter, the thermal conductivity of CNTs shows a downward trend (**Figure 2**) [45]. This is because the increase of diameter inevitably increases the defect content, which leads to more phonon scattering.

CNTs are widely used in various electronic devices due to their high conductivity and chemical stability [46]. For SWCNTs, the specific surface area of SWNTs can reach 240–1250 $m^2 \cdot g^{-1}$, which can generate 180 $F \cdot g^{-1}$ specific capacitance, 20 $kW \cdot kg^{-1}$ power density, and 6.5–7 $Wh \cdot kg^{-1}$ energy density. At the same time, high-temperature heat treatment can reduce the electrode impedance and increase the specific capacitance of SWNTs. The increase of capacitance is considered to be caused by the increase of specific surface area and a large number of 3–5 nm pore distribution [47, 48]. For MWCNTs, they usually have a high specific surface area (about 430 $m^2 \cdot g^{-1}$), a specific capacitance of up to 180 $F \cdot g^{-1}$, a power density of 8 $kW \cdot kg^{-1}$, and an energy density of 0.56 $Wh \cdot kg^{-1}$. CNTs of different shapes (such as direct growth, porous, array, and crimp) have

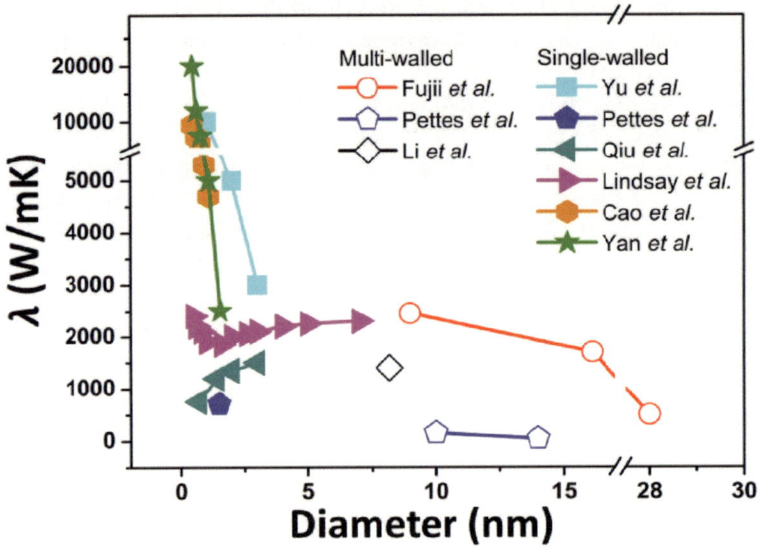

Figure 2.
The relationship between thermal conductivity and diameter of CNTs [45].

been tested as electrodes. The array CNT is the most suitable electrode because of its small internal resistance, good reaction rate, regular gap structure, and stable conductive channel [49–51].

3. Carbon fibers

Carbon fiber is a kind of fiber material with high strength and high modulus. Its carbon content is more than 90%, and CF with carbon content more than 99% is also called graphite fiber, which is mainly composed of disordered graphite microcrystals stacked along the axial direction of the fiber [52]. CF is not only flexible and acid and alkali resistant but also stronger than steel, which makes it an important material for national defense, military industry, and civil use [53].

CF can be classified into polyacrylonitrile-based (PAN-based) CF, asphalt-based CF, viscose-based CF, and gas-phase growth CF according to the source of precursors [52]. As shown in **Figure 3**, according to the basic morphology, it can be divided into filament CF and short CF, wherein filament CF can be woven into two-dimensional CF fabric and three-dimensional CF fabric. Based on the mechanical properties, it can be divided into general CF and high-performance CF which can also be divided into high-strength type (strength >2000 MPa) and high model type (modulus >300 GPa) CF. With the rapid development of aerospace, automobile manufacturing, and sports facilities, the performance of CF has been increased, and the outputs have been improved continuously. Currently, the largest amount of polyacrylonitrile-based CF is used in the real world [54].

3.1 Fabrication of CFs

The industrial production of CF mainly includes polyacrylonitrile-based CF, asphalt-based CF, and viscose-based CF. Among them, the preparation process of viscose-based CF must be graphitized by high-temperature stretching. Because of its complex equipment and technical difficulties, it has not been effectively developed. The production process of polyacrylonitrile-based CF mainly includes two processes:

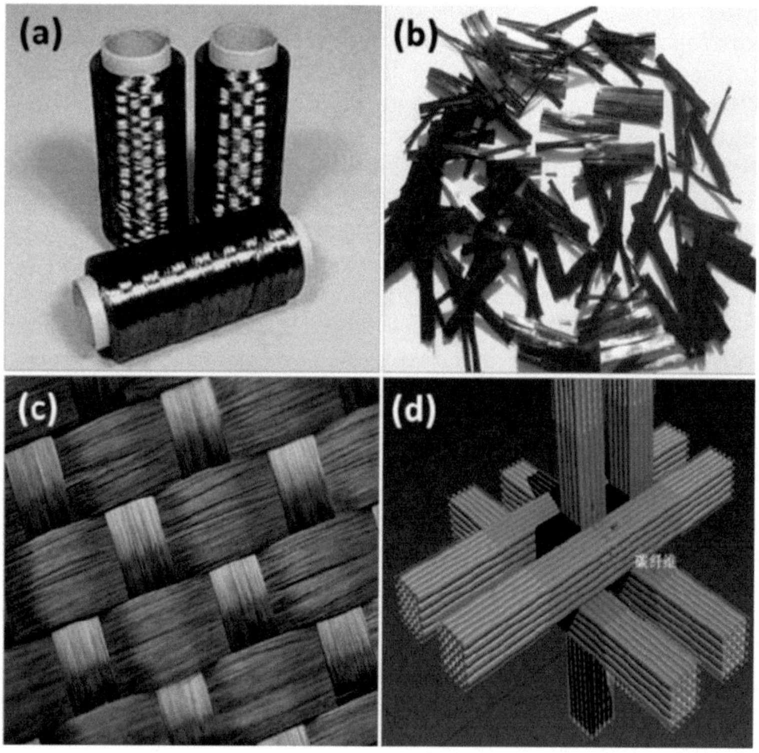

Figure 3.
The pictures of (a) filament CF, (b) short CF, (c) CF cloth, and (d) 3D CF braid.

raw silk production and carbonization. The production process of raw silk mainly consists of polymerization of acrylonitrile monomer, solution defoaming, wire spraying, traction, water washing, oiling, drying, and reeling. Moreover, the carbonization process mainly includes pre-oxidation, low-temperature carbonization, high-temperature carbonization, surface treatment, sizing and drying, winding, and other processes. Note that the pre-oxidation refers to heating the precursor fiber in the air to about 270°C, holding for a period of time, so that the polyacrylonitrile linear polymer will be oxidized, pyrolyzed, cross-linked, and cyclized to form a heat-resistant ladder polymer. In order to prevent melting and deformation of polyacrylonitrile fiber during high-temperature carbonization, the color of polyacrylonitrile fiber gradually changes from white to yellow, then brown, and finally black. The pre-oxidized fiber is carbonized in inert gas with high temperature, and then the cross-linking reaction arises further. With the removal of hydrogen, nitrogen, and oxygen atoms, CF with disordered graphite structure is formed.

The raw material of asphalt-based CF is petroleum asphalt or coal asphalt. The preparation process mainly includes refining, spinning, pre-oxidation, carbonization, or graphitization of asphalt. Among them, mesophase asphalt is a kind of nematic liquid crystal (LC) material composed of disk-shaped or rod-shaped molecules formed by heavy aromatics during heat treatment. The asphalt-based CF prepared by mesophase asphalt is easy to graphitize and usually has a high modulus [52, 55–57].

3.2 Properties of CF

Due to the carbonization and orientation at high temperatures, the carbon atoms of CF are arranged very closely, and the disordered graphite is closely connected. In addition, the diameter of CF is smaller, which can reduce the content of defects, so it has very high mechanical strength and modulus. The tensile strength and modulus of CF can reach 7 and 700 GPa, which are much higher than those of glass fiber and Kevlar fiber. CF can withstand high temperature above 3000°C without contact with air. Therefore, CF has outstanding heat-resistant performance. The higher is the temperature, the greater is the fiber strength. After graphitization, the density of mesophase asphalt-based CF increases, and the carbon content exceeds 99%. Most of the carbon atoms in the fiber form a large area of graphite sheet structure along the fiber axis by SP^2 hybridization, which is very conducive to the phonon vibration. Therefore, the thermal conductivity of the graphite fiber can reach up to 1000 W/mK [45, 58, 59]. It is worth noting that the electrical properties of CF are not ideal, because of the inherent polycrystalline structure and a large number of grain boundaries inevitably formed during the pyrolysis of organic precursors [21].

4. Graphene-based fibers (GBFs)

Graphene is a two-dimensional (2D) crystalline sheet with a monolayer of carbon atoms densely packed in an SP^2-bonded honeycomb lattice and can be considered as a single layer of the graphitic film in graphite. Thus, graphene is the thinnest nanomaterial known [60, 61]. As shown in **Figure 4**, the length of carbon–carbon bond in graphene is about 0.142 nm; all carbon atoms are connected with three surrounding carbon atoms by σ bond; the remaining P electron orbit is perpendicular to the plane of graphene to form delocalized π bond because π electron can move freely in the plane, rendering graphene holding excellent electrical properties [62, 63].

Since graphene was found in 2004 [61], because of its unique physical and chemical characteristics, such as extraordinary thermal conductivity [64], mechanical strength ($\sigma_{int}^{2D} = 42 \pm 4(N \cdot m^{-1})$) [65], and fast electron mobility ($\mu \approx 10{,}000$ $cm^2 \cdot V^{-1} \cdot s^{-1}$) [66–70], it has aroused great interest. Due to the oxygen-containing functional groups, graphene materials obtained from chemical methods such as graphene oxide and reduced graphene oxide (rGO) are highly maneuverable and reactive, which further inspires a wide range of research enthusiasm in preparation, chemical modification, and well-controlled assembly of advanced and macroscopic structures for various device applications [71–77]. To this end, graphene-based 3D aerogels (GBAs), 2D membranes (GBMs), and 1D fibers (GBFs) have been developed. Among them, GBAs hold the current world record for the lightest material,

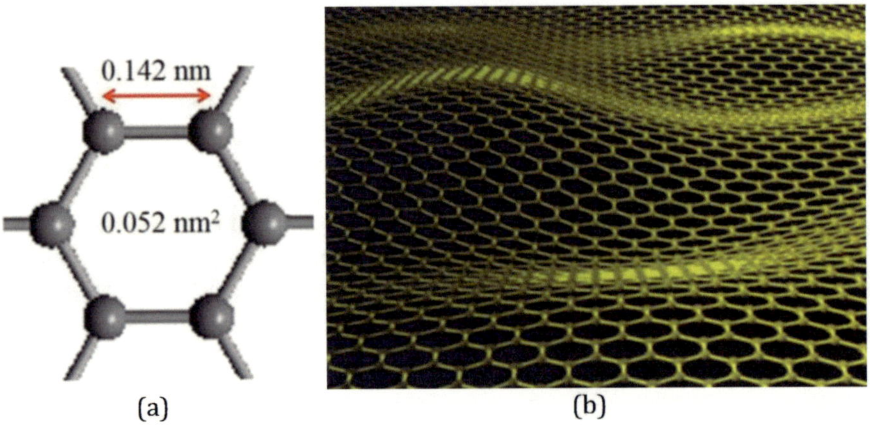

Figure 4.
(a) Schematic diagram of a honeycomb crystal lattice of graphene, and (b) a single-layer suspended graphene sheet exhibits intrinsic microscopic roughening.

with a density of 0.16 mg · cm^{-3} [78], and have demonstrated good capability in the removal of spilled oils [78, 79]. GBMs, which are usually fabricated by infiltration or CVD, have also found to have various applications in the field of energy storage and conversion [80]. Compared with GBAs and GBMs, GBFs possess not only outstanding mechanical property and high conductivity but also good valuable flexibility that can be curved, knotted, and even woven into flexible conductive fabric, which are considered capable of improving the practical applications of GBFs. The development of high-performance GBFs could inspire more engineering applications of graphene. However, assembling microscopic graphene sheets into 1D fiber remains as an unusual challenge because of the irregular shape and size and the movable stacked layers of graphene sheets compared with the highly tangled CNT assemblies [63]. Nevertheless, the assembly of graphene sheets into macroscopic fibers has attracted wide interest due to the lightweight, lower cost, shareability, ease of functionalization, and practical importance of GBFs in contrast to CNTs and CF. Beyond that, 1D GFs with mechanical flexibility is particularly important for wearable textile devices and can serve as the building blocks for constructing 2D and 3D macroscopic architectures for various applications.

4.1 Fabrication of GBFs

At present, the manufacturing methods of GBFs are mainly influenced by traditional synthetic fiber production methods, including melt spinning and solution spinning [63]. However, due to the high-temperature stability of graphene, its melting temperature is even higher than that of fullerene and carbon nanotubes. Therefore, melt spinning is not the choice for manufacturing GBFs, while solution spinning is [81, 82]. Solution spinning mainly includes wet spinning, dry jet wet spinning, and dry spinning. In addition to these traditional solution spinning methods, some new methods, including electrophoresis, template hydrothermal method, and chemical vapor deposition-assisted assembly, have been developed recently. In this part, the common methods of preparing GBFs will be introduced in detail.

4.1.1 Wet spinning

Wet spinning is one of the main methods to prepare chemical fiber. The important step is to prepare a spinning solution. Because graphene is not easily dispersed in water or other organic solvents, it is difficult to prepare a spinning solution, so it is not possible to prepare fibers from graphene by wet spinning [83–85]. As an important precursor of graphene, graphene oxide can be well dispersed in polar solvents (such as water), so it is expected to prepare fibers by wet spinning [86]. The steps of preparing GBFs by wet spinning are as follows: first, GO dispersions are injected into a stable aqueous solution to form GO spinning dope and then injected into the coagulation bath to form a gel-like fiber to prepare GO dope. After solidification for a period of time, GO fiber can be obtained by extracting colloidal fiber and drying, and then GO fiber can be reduced to produce GBFs, as shown in **Figure 5**. An rGO fiber can be further produced by reducing the GO fiber when needed [86, 87]. To ensure uniform and continuous formation of gelatinous fibers, the fibers after solidification should be kept at a certain speed. They can be drawn through a rotating bath or using a collecting unit, as shown in **Figure 5**. The highest strength rGO fiber is made by the method shown in **Figure 5a**. This method includes an easy spin of a small amount of fiber, but it lacks accurate control of fiber moving speed. In contrast, the method shown in **Figure 5b** can provide constant traction and determined moving speed to synthesize fibers, so the method is more suitable for producing fibers with accurate tensile ratio and good scalability [88].

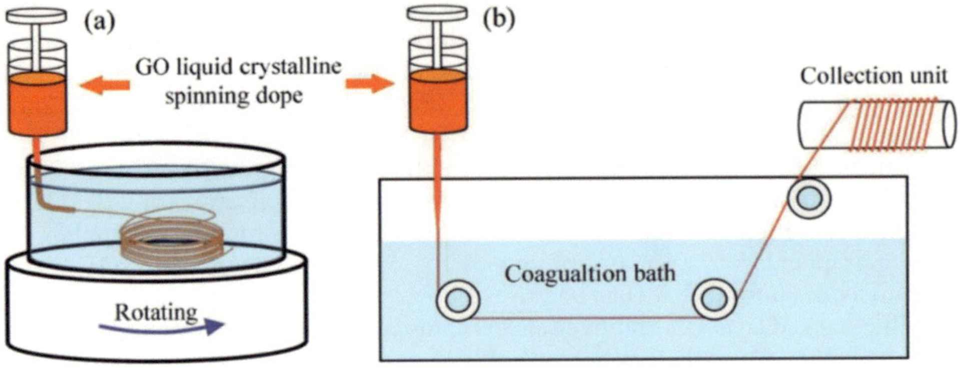

Figure 5.
(a) The synthesis of graphene oxide fiber by wet spinning in rotating coagulation bath and (b) collection unit [86, 87].

Zhen et al. prepared liquid crystal GO aqueous solution for the first time in 2011, taking NaOH/methanol solution as coagulation bath, obtaining GO fiber through wet spinning, and then reducing GO fiber in hydroiodic acid to produce GBFs. This method can make GO sheets form liquid crystals, which can enhance the strength and flexibility of GBFs. The tensile strength of the fiber is 140 MPa, and the conductivity is $2.5 \times 10^4\ S \cdot m^{-1}$ [86]. Then, Zhen et al. further tried to increase the lamella of the raw material GO, using N,N-dimethylformamide (DMF) as the solvent, acetone, and ethyl acetate mixture as the coagulation bath. After that, the mechanical ability of GBFs was improved by spinning drafting and high-temperature treatment at 3000°C, making its strength reach 1.45 GPa [89]. On the other hand, the conductivity of GBFs can be improved by ion doping, and the conductivity of potassium doped GBFs can reach $2.24 \times 107\ S \cdot m^{-1}$ [90]. In addition, Shaohua et al. prepared non-liquid crystal GO aqueous solution to achieve a high concentration of spinning solution to improve the fiber yield. The concentration of a spinning solution can reach 2%, and then GBFs were obtained through a similar wet spinning process and reduction by hydrocodone. The mechanical and electrical properties of the fiber were 208 MPa and $1.53 \times 103\ S \cdot m^{-1}$ [91], respectively.

In addition to GBFs, graphene composite fibers can also be prepared by wet spinning, so as to effectively improve the fiber performance and expand the application field. Conducting polymer monomers are polymerized in situ during spinning to prepare composite fibers [92, 93], or oxides or other materials are added directly into the spinning solution to increase the capacity of fiber-shaped supercapacitors [94, 95]. Wujun et al. used GO to disperse the water-insoluble activated carbon in the aqueous solution, spinning and reducing to obtain graphene/activated carbon composite fiber. The fiber has a specific surface area of $1476.5\ m^2 \cdot g^{-1}$ and a capacity of $43.8\ F \cdot g^{-1}$ [96]. Similarly, the graphene/manganese dioxide composite fiber can be spun by a similar process, and the capacity of the supercapacitor can reach $66.1\ F \cdot cm^{-3}$ [97]. In addition to inorganic materials, GO and polyvinyl alcohol (PVA) also have good compatibility. Adding sodium hydroxide to non-liquid crystal GO aqueous solution for a pH = 11. Then, adding PVA can significantly increase the affinity between fiber and electrolyte [98]. Similarly, the surface of the fiber with a large number of hydroxyl groups can significantly increase the hydrophilicity and strength of the fiber, which is caused by a large number of oxygen-containing functional groups on its surface [99]. Mochen et al. developed a method to improve the strength of graphene fiber. They spun GO and phenolic resin together. After carbonization under the condition on 1000°C, the C–C covalent bond was formed between graphene sheets, and the fiber strength reached 1.45 GPa [100].

The tensile strength and elongation at break of graphene fiber with 10% phenolic resin are 1.45 GPa and 1.8%, respectively, which are better than most GBFs reported before. The increase of strength, toughness, and elongation can be attributed to the formation of a C–C bond between the graphene sheet and phenolic carbon, which provides sliding space for the graphene sheet before fracture. Yang et al. developed a simple but effective method for continuous manufacturing of neat, morphologically defined, graphene-based hollow fibers (HFs) with coaxial capillary spinning strategy. As shown in **Figure 6**, the preparation method of GO-HFs is to use coaxial capillary spinneret to spray silk in 3 mol $\cdot$ L^{-1} KCl methanol solution and use compressed air to replace the internal fluid of KCl/methanol solution to successfully prepare GO-HFs with necklace structure (nGO-HF). Experiments show that nGO-HF has a large elongation of about 6% when it breaks, which indicates that nGO-HF has a strong ability to bear compression, which is caused by the elastic deformation of hollow microspheres [101]. Therefore, the physical properties of GBFs can be controlled by adjusting the spinning conditions.

4.1.2 Dry spinning

In the dry spinning of GBFs, GO dispersion (mainly dispersed in water) is also used as a spinning assistant rather than a coagulation bath. Instead, the GO dispersion is injected and sealed in a pipe, and the GO dispersion is precipitated in the form of gel state fiber at high temperature by heating or chemical reduction, and then dried rGO fibers can be obtained by further solvent removal. GO dispersions are considered to be colloids with large-size dispersants [102–104]. The study of Dong et al. and Yu et al. shows that high temperature can promote the rapid movement of GO dispersant and increase the possibility of collision and precipitation of GO plate. At the same time, high temperature or chemical reduction can also separate the

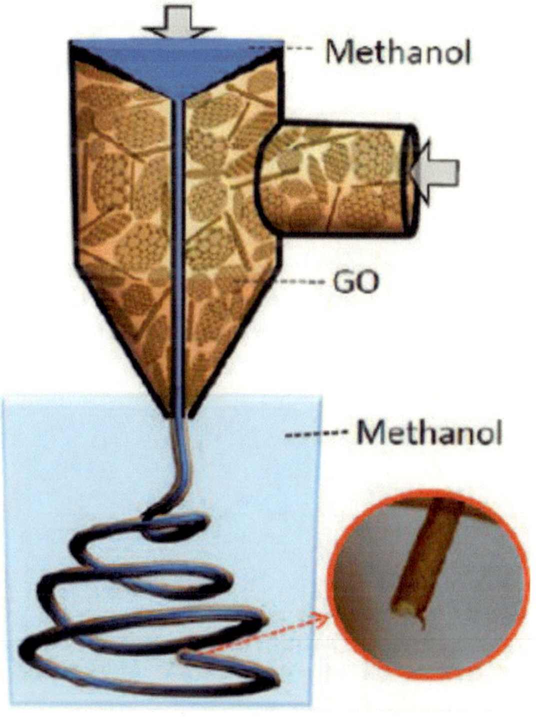

Figure 6.
Schematic of the setup that used a dual-capillary spinneret to directly spin GO-HFs [102].

oxygen-containing groups in GO and reduce the zeta absolute potential of GO dispersion. Finally, due to the lack of sufficient electrostatic repulsion, GO sheets' precipitate is assembled into fibers. The fibers in the gelatinous state expand in the solvent, but the diameter of the fibers can be reduced by about 80% after drying [105, 106].

In the process of dry spinning, the precipitate of GO sheet under the condition of 220–230°C is actually a solvothermal process (the water is used as the solvent to disperse GO). The process flow is shown in **Figure 7** [107], and the fibers made in this way are actually rGO fibers. It is reported that 27% of oxygen in GO can be removed at 180°C and most hydroxyl, epoxy, and carboxyl groups begin to separate at 200°C [108, 109]. Therefore, GBFs synthesized by the hydrothermal method has considerable conductivity, without post-reduction treatment [105, 110]. In addition to the hydrothermal method, Jihao et al. also use the chemical reduction method of CO to produce rGO fiber. First, the GO and vitamin C (VC) solution was injected into the polypropylene (PP) tube, then heated to 80°C, and kept for 1 h, while GO was reduced and assembled into gel-like rGO fibers. After extraction and drying, the fiber diameter decreased by 95–97%, which was due to the shrinkage of the fiber due to the removal of moisture. Finally, the conductivity of the rGO fiber is about 8 $S \cdot cm^{-1}$ [111].

In the dry spinning process, GO dispersion does not necessarily exist in the form of liquid crystal [106]. The randomly dispersed low-concentration GO dopes (8 $mg \cdot ML^{-1}$) composed of small-diameter GO (d = 0.5–4 μm) can also be used for the synthesis of GBFs [112, 113]. Compared with the wet spinning method, the preparation of graphene fiber by dry spinning does not need other auxiliary reagents, only needs high-temperature reduction or chemical reaction to get GBFs, and can get any shape of GBFs by pre-setting template, even hollow structure GBFs [105, 106, 112]. It is worth noting that after reduction, the strength of GO fiber obtained by dry spinning is lower than that by wet spinning, but its toughness is as high as 19.12 $MJ \cdot m^{-3}$, which is expected to become a green processing method of GBFs in the future [107].

4.1.3 Dry jet wet spinning

Dry jet wet spinning is another important spinning method of conventional synthetic fiber. The results show that PAN-based carbon fibers can be spun with high concentration coating by this method and the mechanical properties of the fiber are better than that of wet spinning [52, 114]. Shayan et al. use dry jet wet spinning to improve the strength of the fiber. The existence of the air layer effectively reduces the speed gradient of the spinning liquid from the spinneret to the coagulation bath, so that the fiber has a better arrangement. However, if the air layer is too long, it will affect the tensile property of the fiber and control the diameter of the needle and the distance of the air layer. Then, the high-strength GBFs with circular cross section can be spun (**Figure 8a**) [115].

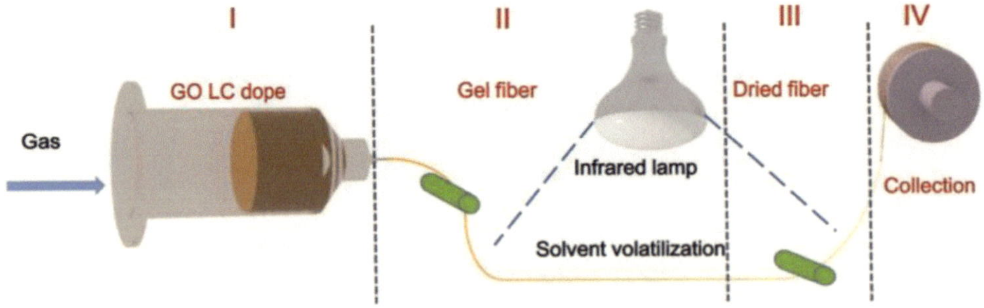

Figure 7.
Schematic illustration of the dry spinning process with a concentrated organic dispersion of GO [108].

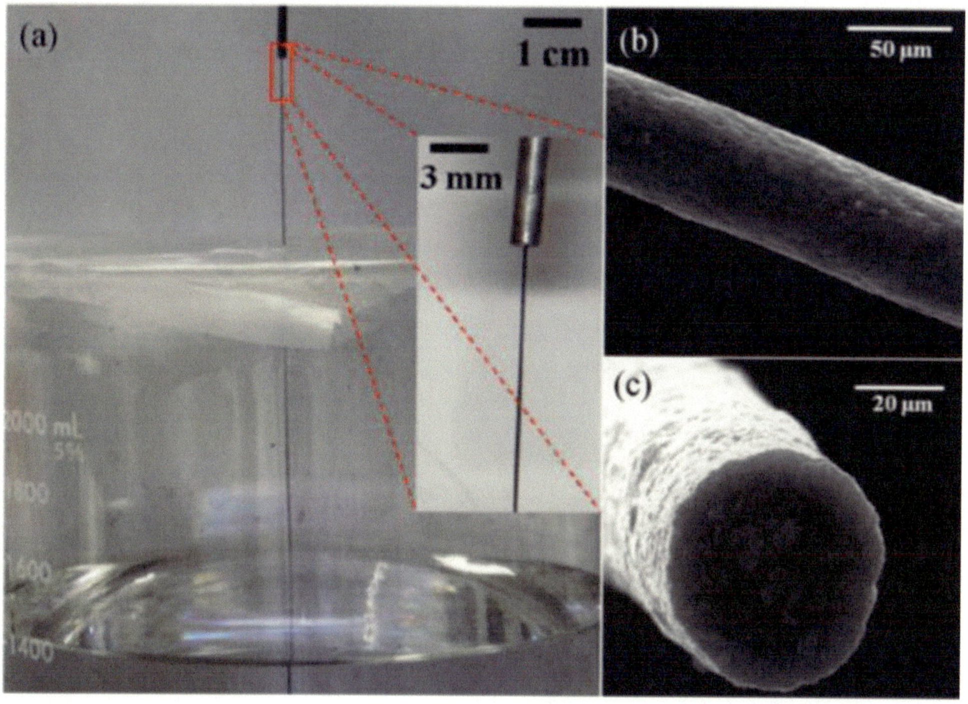

Figure 8.
(a) A digital photo showing the setup for dry jet wet spinning of GBFs [115]. (b) and (c) Surface and cross-sectional images of a GBFs [103].

Figure 8b and **c** shows the surface and cross-sectional images of dry jet wet spinning fiber, which indicates that the GBFs with smooth surface and circular cross section can be produced by dry jet wet spinning with proper solvent coalescent pair (chlorosulfonic acid and diethyl ether), which is not realized in both wet spinning and dry spinning.

4.1.4 Chemical vapor deposition-assisted assembly

In the production of graphene made by chemical vapor deposition, the composition of graphene can be easily changed by changing the composition of the gas phase. Xinming et al. reported in 2011 a method of self-assembly of two-dimensional CVD grown films into one-dimensional GBFs in ethanol, acetone, and other organic solvents through the change of surface tension. The resulting fibers have a high conductivity of about 1000 $S \cdot m^{-1}$ [116]. Seyed et al. reported another method of film assembly. They first scraped and coated the GO dispersion into multiple film strips, dried and twisted it to get GO fiber, and then put it through thermal reduction to get GBFs. The GO fiber made by this method has high elongation at break (8.3–78.3%) and excellent fracture toughness (1.3–17.4 $J \cdot m^{-3}$), but its strength is low (9.7–85.9 MPa) due to many defects in the fiber section [117]. Jiali et al. also developed a method for preparing GBFs by film shrinkage, as shown in **Figure 9**. First, graphene was produced on copper foil by CVD with methane as a carbon source. In order to obtain a complete and independent graphene film, a layer of polymethyl methacrylate (PMMA) is spin-coated on the surface of graphene. The copper foil is etched with 1 M ammonium persulfate solution, and the PMMA layer is washed off with acetone to obtain the laminated graphene film. Second, the film is pulled out of the solution with tweezers to shrink to form GBFs with uniform diameter [118]. The graphene film can be directly used to prepare GBFs by film

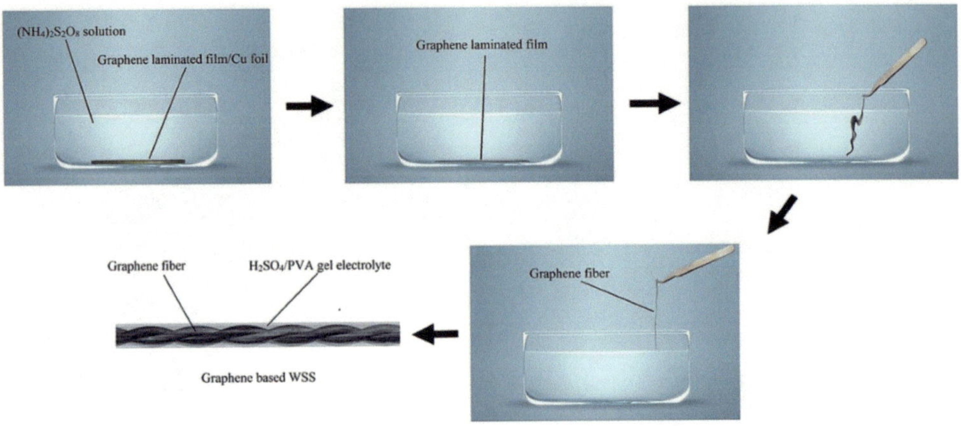

Figure 9.
Schematic illustration of GBFs prepared by film assembly [118].

shrinkage method, and the obtained fiber generally has more pores. However, the CVD method needs a lot of instrument investment and strict gas conditions, and the cost is high, so it is difficult to promote.

4.1.5 Templated hydrothermal method

Zelin et al. reported a template hydrothermal method to prepare GBFs. GO dispersion was injected into the stripping pipe, sealed at both ends, and then heat-treated in water at 230°C for 2 h to form continuous GBFs. The structure of GBFs can be adjusted by controlling the concentration of GO dispersion and the inner diameter of a glass tube. The graphene fiber has a porous structure, has a density of only $0.23\ g \cdot cm^{-3}$, and has a good flexibility [105]. Yunming et al. used a simple low-temperature-induced self-assembly method to synthesize GBFs. They mixed GO and ascorbic acid evenly and sealed them in a specific straight glass tube. They carried out the hydrothermal reaction at 90°C and 120°C, respectively, until the fiber was completely formed and then obtained GBFs with layered porous structure. Its conductivity can reach $1.3 \times 10^4\ S \cdot m^{-1}$. After heating, it has excellent mechanical properties and can be easily woven into the spinning products [119]. Lizhi et al. further developed on the basis of previous methods, and the specific preparation process is shown in **Figure 10**.

First, the dispersion of GO is sprayed into liquid nitrogen through a spout to prepare a layer bridging GO dispersion - interconnected graphene oxide ribbons (IGOR). Then, a certain concentration of GO dispersion is uniformly mixed with IGOR dispersion and injected into a quartz capillary with an inner diameter of 0.4 mm. The two ends are sealed, heated at 230°C for 2 h, and finally dried in air for 12 h. The GBFs show higher strength and toughness [120].

In order to increase the length of GBFs prepared by the hydrothermal method, Dingshan et al. improved the above methods. The authors replaced the brittle glass tube with the flexible and high-temperature-resistant fused silica capillary column, injected the GO dispersion containing ethylenediamine into it, and sealed it. After that, they put it in the furnace at 220°C for 6 h, extruded it with nitrogen to form the fiber, dried it, and collected the long enough GBFs [106]. Although the GBFs with porous structure can be prepared by the hydrothermal method, it is difficult to achieve continuous production because of the need of closed space and long reaction time.

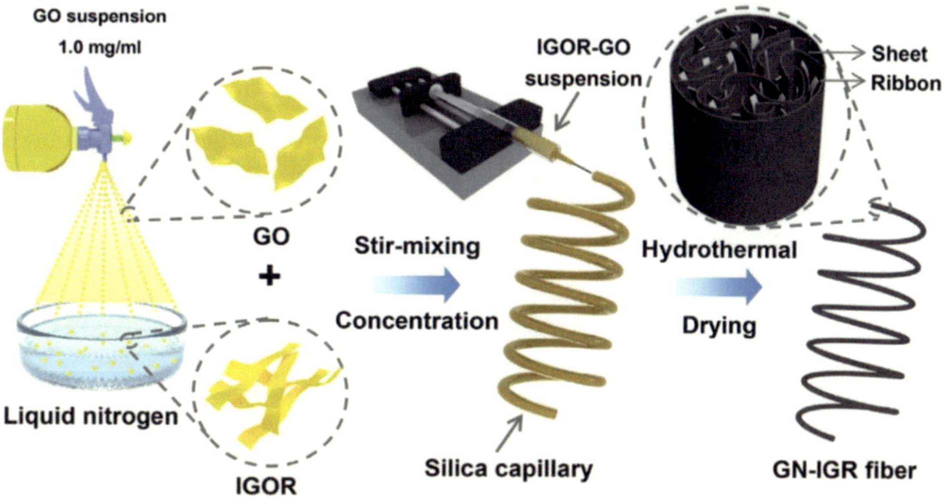

Figure 10.
Schematic illustration of graphene hybrid fibers prepared by hydrothermal method [120].

4.1.6 Conductive substrate-assisted spontaneous reduction and assembly

The conductive substrate-induced spontaneous reduction and self-assembly of GO generally proceeds by putting metal substrates (e.g., Al, Fe, Cu) into GO solution for GBF preparation. As shown in **Figure 11**, Junjie et al. take copper wire as the substrate and adopt the three-electrode method to make the GO sheet continuously deposit on the surface of copper wire under the double induction of electrochemistry and template. Both GO and copper are simultaneously restored. Then, they etch and remove the copper wire in the $FeCl_3$ solution to obtain the graphene hollow fiber with an oriented structure. The controllable preparation of the hollow fiber can be realized by controlling the diameter, length of the substrate, and the time of electrochemical deposition. The graphene hollow fiber has excellent flexibility and conductivity and can be used as the electrode material of supercapacitor [121].

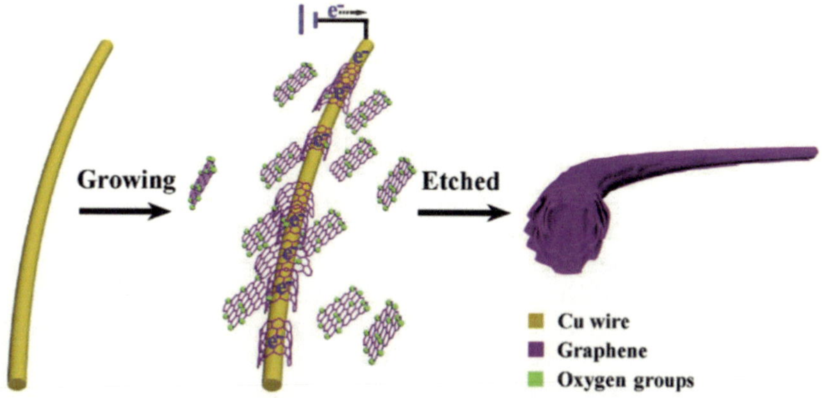

Figure 11.
Scheme of spontaneous reduction and assembly of graphene hollow fiber on active metals substrates [121].

4.1.7 Electrophoresis self-assembly method

The electrophoretic phenomenon occurs in a colloidal solution because charged particles can move under the action of electric field. Lianlian et al. developed a method for preparing GBFs with electrophoretic self-assembly. The graphite probe was used as a positive electrode to invade the GO dispersion. Under constant potential, the graphite probe was extracted slowly and uniformly, and self-assembled GO fibers were formed at the tail of the cathode. After drying and heating, GBFs with a smooth surface and circular cross section can be obtained [122]. Because the electrode moving speed is only 0.1 $mm \cdot mm^{-1}$, it takes 1 week to get 1-m-long fiber. The yield of GBFs obtained by this method is too low to scale production.

4.2 Applications of GBFs

Thanks to graphene's superior electrical, mechanical, and thermal properties and good flexibility, GBFs have great potential in sensor, energy storage, energy conversion, and other fields.

4.2.1 Sensor

With the continuous development of flexible equipment, intelligent devices, including electricity, humidity, force, and temperature, can rapidly make structural changes in the environment and be increasingly concerned by people. The GBFs shows excellent performance in this regard.

Zhao et al. successfully developed a graphene-based multifunctional optical fiber sensor, which can respond to three different stimulations. They deposited GCN on GF (GF and GCN) and twisted it with another GF to form a double helix GBFs. In the twisted structure, the contact interface of the two fibers has a sandwich-like graphene/GCN/graphene structure. Under different external voltage controls, GF and GCN can show three different stimulus modes. Each mode can respond to temperature fluctuation, mechanical interaction, and humidity change and has a high sensitivity to specific stimulation [123]. Yanhong and his team electroplated polypyrrole on half of the surface of GBFs, which changes the current transmission rate on both sides of the fiber. With different types of current, the fiber has different bending states. The prepared electric GBFs are expected to be applied in the multi-arm tweezers and mesh driver [124]. Chunfei et al. used twisted GBFs to realize temperature sensing. With the increase of temperature, the fiber resistance decreases. This is mainly due to the transition of semiconductor characteristics between graphene sheets. The fiber has similar sensing characteristics for temperature under different stretching conditions and has a wide application prospect [125].

In addition, GO fiber is partially restored by laser method, which is sensitive to humidity. By changing the position, the fiber can be transformed into various shapes. Taking advantage of the hydrophilic characteristics of GO in a humid environment, the distance between sheets is increased, while graphene is non-hydrophilic. Hence, the bending degree of the fiber changes with the humidity. Meanwhile, the fiber is woven into fabric shape, which still has sensitive response performance [126]. After twisting the spinning GO fiber, the twisted fiber will rotate repeatedly as the humidity changes periodically. When the humidity increase, a large number of oxygen-containing functional groups on the surface of GO will absorb water, and the distance between layers will increase. Otherwise, the distance between layers will decrease. A magnet is added at the lower end of the fiber to prepare a humidity sensing electric motor. The speed of the motor reaches 5190 $r \cdot min^{-1}$. The motor can convert the change of environmental humidity into electric energy and realize the collection of energy [127].

The GBFs and the GBFs coated with a layer of carbon nitride on the surface are wound together. The middle carbon nitride layer is equivalent to a buffer layer. Its conductivity is related to the layer spacing. With the pressure increase, the distance decreases and the conductivity is, in turn, to increase, which can realize the stress sensing [123].

4.2.2 Energy storage

4.2.2.1 Lithium-ion batteries

With the development of science and society, a portable energy storage device is becoming smaller and more flexible. Lithium-ion batteries are a new type of energy storage device, which has the advantages of high energy density, environmental friendliness, long cycle life, and high working voltage. However, the traditional LIBs cannot meet the needs of wearable electronic devices due to its large usage, rigidity, and weight. Therefore, it is necessary to develop new batteries with small volume, lightweight, and high flexibility. GBFs maintain the unique characteristics of the graphene nanosheet. When GBFs are used in the fiber lithium battery, it can realize the series connection with flexible electronic devices and drive them to work stably, achieving high energy density and holding a good commercial prospect [128, 129].

Jung et al. of the Korea Institute of Chemistry used pure GBFs as the negative electrode material of lithium-ion batteries. The battery circulates 100 times in the range of 0.005–3 V under the current density of $100\ mA \cdot g^{-1}$, and the capacity is still $224\ mAh \cdot g^{-1}$ [130]. Minsu et al. obtained hollow GBFs by coaxial spinning and increased specific surface area and active site, and its capacity remained $196\ mAh \cdot g^{-1}$ in the range of 0.005–1.5 V for 100 cycles under the current density of 0.2C [131]. Due to the low capacity of pure GBF battery, Jong et al. added MnO_2 active material in graphene; the addition of MnO_2 increased the distance between graphene sheets and gave lithium-ion fast transfer channel. Moreover, the battery made by MnO_2 coating of graphene has good cycle stability, and the cycle capacity of 100 times remained $560\ mA \cdot g^{-1}$. Minsu et al. filled the inner space with Si/Ag nanoparticles, and the outer graphene well controlled the volume expansion of the inner silicon during charging and discharging, providing a smooth electronic channel. Compared with the simple mixing process, it has better cycle stability and rate performance, and the capacity of 100 cycles remains $766\ mAh \cdot g^{-1}$ [131].

The GBFs prepared by the above method have low strength, and it is difficult to form a macroscopical fiber battery. In one report, a fiber battery electrode comprised of 2D/2D layered titania sheets/rGO sheets (titania/rGO) composites was prepared through wet spinning method [132]. By assembling the cathode of titania/rGO fiber with the anode of lithium wire in parallel, a fiber-shaped half-cell was fabricated. This hybridized fiber electrode had an ordered stacking structure, high linear density of active materials, and abundance of exposed active sites, which endows the fiber electrode with prominent mechanical flexibility combined with excellent battery performances of high linear capacity of $168\ mAh \cdot g^{-1}$, good rate capability, and outstanding cyclic behavior. Woon et al. used wet spinning to construct graphene/carbon tube/sulfur electrode as positive material of Li-S battery. Graphene has high conductivity and can transfer electrons rapidly. Meanwhile, GO fiber as a matrix can obtain light fiber with certain mechanical strength for wearable equipment, as shown in **Figure 12a** and **b** [133].

Compared with wet spinning, the diameter of the nanofiber film obtained by electrospinning is smaller. As the electrode material of lithium battery, it can significantly reduce the migration distance of lithium-ion and increase the specific surface area of the electrode material and improve the electrochemical performance

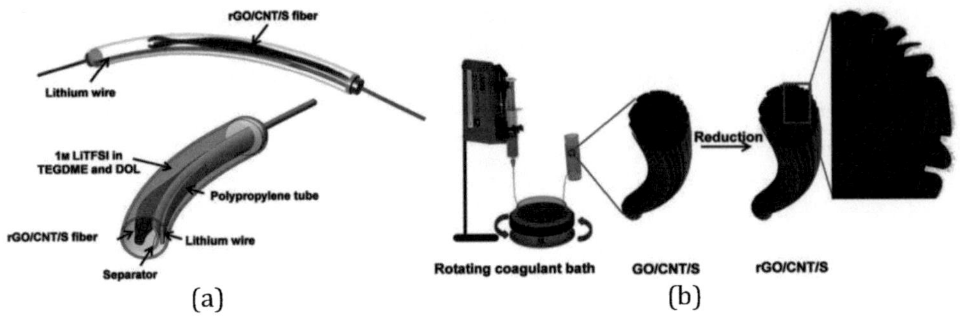

Figure 12.
(a) Schematic of fiber-shaped lithium-ion battery. (b) Schematic illustration of synthetic route of rGO/CNTs/S fiber [133].

of the battery [134–136]. Xiaoxin et al. obtained the Si-graphene-C structure which is similar to the coronary artery based on bionics. Graphene can effectively control the volume expansion of Si, and high conductivity is also conducive to the rapid transfer of ions. Meanwhile, the inclusion of graphene also avoids direct contact between Si and electrolyte and avoids the formation of a large number of SEI films. After 200 cycles, the capacity retention rate is still 86.5% [137]. Jian et al. continued to wrap a layer of graphene outside SnO_2 and GO nanofibers with a double-layer protection method to inhibit the volume expansion and agglomeration of active materials. This method is applicable to almost all oxide and graphene nanofiber electrodes obtained by electrospinning, with good universality [138].

At present, there are few researches on the application of GBFs in LIB and the assembly of woven fiber batteries. Compared with the traditional button batteries, the assembly process of GBFs is relatively complex, so it is unable to achieve continuous production.

4.2.2.2 Supercapacitor

In addition to the application in LIB, GBFs are also widely used in the field of supercapacitors. Supercapacitor, also known as a double electric layer capacitor or electrochemical capacitor, is a new energy storage device that uses the rapid adsorption–desorption of electrolyte ions with electrode materials or the reversible oxidation–reduction reaction on the surface of electrode materials to realize electric energy storage [139, 140]. With the continuous development of wearable devices, flexible supercapacitors have become the preferred energy source for various electronic devices due to their fast charge and discharge ability and long cycle life. Among them, fiber supercapacitors have attracted much attention due to their lightweight, small size, high flexibility, and good wearability. GBFs have excellent conductivity and super high specific surface area, so it has been widely used in the field of fiber supercapacitor [141].

Chen et al. prepared pure GBFs with a non-liquid crystal method and further assembled the fibers into flexible supercapacitors. The capacitance of the supercapacitor is $39.1\ F \cdot g^{-1}$ when the current density is $0.2\ A \cdot g^{-1}$. At the same time, it is found that the electrochemical performance of GBFs can be greatly improved by immersing it into 6 M KOH for 10 min before the electrochemical performance test. At the current density of $0.2\ A \cdot g^{-1}$, the specific capacitance is $185\ F \cdot g^{-1}$ ($226\ F \cdot cm^{-3}$), and the energy density is $5.76\ Wh \cdot kg^{-1}$ (power density is $47.3\ W \cdot kg^{-1}$) [91]. The capacitor has good toughness and can be woven into fabric and light LED after charging. Hu and Zhao integrated two electrodes (the upper and lower part of rGO) and separator (the middle part of GO) into the GO optical fiber, as shown in **Figure 13a**, and made a kind of all-in-one fiber graphene supercapacitor (rGO-GO-rGO) without any adhesive. The diameter of the rGO-GO-rGO fiber is 50 μm, and the rGO part is about 1/4 of the fiber width. The rGO-GO-rGO fiber supercapacitor shows remarkable mechanical flexibility, which can bend to various curvature while maintaining high capacitance (**Figure 13b**) [142, 143].

At present, the specific capacitance of pure GBFs is far less than the theoretical capacitance of graphene. How to improve the capacitance of GBFs is still a big challenge. Currently, an effective method that has been proven and widely used is the hybridization strategy, including doping and compounding with other substances.

Doping increases the active region on the surface of graphene and further improves its catalytic activity for a redox reaction. Among all kinds of atom doping, nitrogen atom doping is the most common. Doping nitrogen atoms with extra valence electrons into graphene will introduce new energy into the low energy region of the carbon conduction band. The introduction to this new energy can improve the catalytic activity and electrochemical performance of graphene materials. Yunzhen et al. extruded the GO dispersion into the substrate of hydroxylamine ethanol solution as a network, dried it, and heat it to obtain the nitrogen-doped rGO network fabric. Then, the PT foil was used as the collector to assemble the supercapacitor. The specific capacity was 188 $F \cdot g^{-1}$ when the scanning rate was 5 $mV \cdot s^{-1}$ in 25% KOH electrolyte. When the scanning rate was increased to 1 and 10 $V \cdot s^{-1}$, the specific capacity was kept at 74.2 and 48.4%, respectively, showing very excellent rate performance [144]. Guan et al. constructed nitrogen-doped porous GBF supercapacitors with high energy density output, large-scale weaving, and flexible wearable application prospects by means of self-assembly of the liquid–liquid interface and molecular functional doping pore formation in the micro-reaction system. The area-specific capacitance of the fiber supercapacitor prepared by this method is as high as 1132 $mF \cdot cm^{-2}$, which has excellent cycle stability and bending durability [145].

Graphene can be compounded with other carbon nanomaterials, conducting polymers, metal oxides/sulfides, and other materials to form graphene composite fibers. The high specific capacitance of the additives can be used to improve the electrochemical performance of the composite fibers.

Yu et al. constructed a graphene/CNT composite fiber. Due to the high conductivity of CNTs, the conductivity of the composite fiber can reach 102 $S \cdot cm^{-1}$, and the specific surface area can reach 396 $m^2 \cdot g^{-1}$. The volume-specific capacitance of the fiber electrode is 305 $F \cdot cm^{-3}$, and the mass-specific capacitance is 508 $F \cdot g^{-1}$ [106]. Yuning et al. mixed GO and pyrrole monomers as spinning solution and extruded

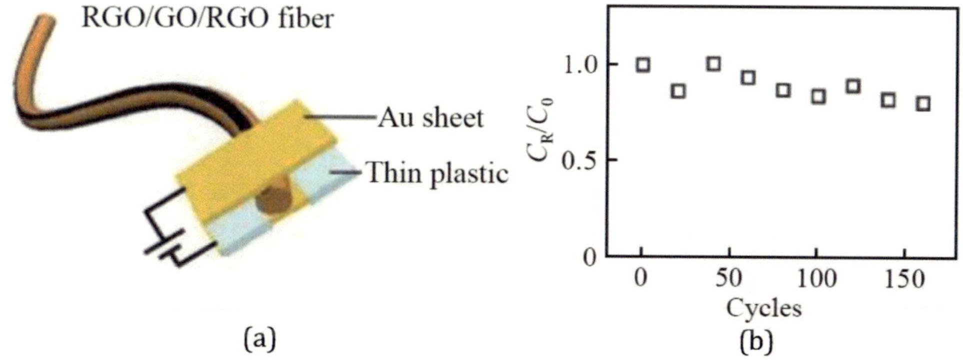

Figure 13.
(a) Scheme of supercapacitor supported by two electrodes. (b) Capacity decrease with increasing bending cycles [142].

them into $FeCl_3$ solution to solidify and polymerize pyrrole in situ, and the PPy/GO composite fiber was obtained after reduction by hydroiodic acid. The fiber has a skin core structure, and its capacitance performance is greatly improved compared with pure rGO fiber. The area-specific capacitance is 107.2 $mF \cdot cm^{-2}$ (73.4 $F \cdot g^{-1}$), and the energy density is between 6.6 and 9.7 μ $Wh \cdot cm^{-2}$ [146]. Bingjie et al. synthesized the graphene/molybdenum disulfide composite fiber electrode with the one-step hydrothermal method. The electrode has a new intercalation nanostructure, which effectively combines the excellent conductivity of the graphene sheet layer with the high pseudocapacitance of molybdenum disulfide. The final assembled fiber-like super electric container shows a volume-specific capacitance of up to 368 $F \cdot cm^{-3}$ [147]. Qiuyan et al. overcame the problem of poor interaction between MXene layers and prepared MXene/graphene composite fiber. The orientation distribution of MXene sheets among GO liquid crystal templates realized high load (95 w/w%). The composite fiber shows excellent conductivity (2.9×10^4 $S \cdot m^{-1}$) and ultrahigh-volume-specific capacitance (586.4 $F \cdot cm^{-3}$), far exceeding the value of pure GBFs [148].

In addition, the structure optimization of GBFs is also an effective way to improve the performance of GBF supercapacitor, which mainly lies in the improvement of specific surface area and the regulation of the layer arrangement structure. The porous GO fiber reported by Seyed et al. in 2014 was transformed into porous rGO fiber after thermal reduction at 220°C, as shown in **Figure 14**.

The specific surface area of the fiber is 2210 $m^2 \cdot g^{-1}$, and the conductivity is about 25 $S \cdot cm^{-1}$, and the specific capacity of the fiber is 409 $F \cdot g^{-1}$ when the current density is 1 $A \cdot g^{-1}$. The specific capacitance of 56 $F \cdot g^{-1}$ still exists when the current density is increased to 100 $A \cdot g^{-1}$ [117]. Chen et al. used cellulose nanocrystals (CNC) to adjust the structure of GBFs. CNC nanorods can not only improve the serious accumulation of graphene sheets in GBFs but also inhibit the possible bending and folding of graphene sheets in the process of fiber-forming, so as to form ordered nanopore structure. The composite GBFs were assembled into a supercapacitor with a conductivity of 64.7 $S \cdot cm^{-1}$ and a specific capacitance of 208.2 $F \cdot cm^{-3}$, which has excellent electrochemical performance [99]. In addition, they also use graphene hollow fiber prepared by the electrochemical method as the electrode of fiber-like supercapacitor [121], and the additional inner surface of hollow fiber can provide more contact area with electrolyte. Under the current density of 0.1 $A \cdot g^{-1}$, the specific capacitance of the assembled solid-state supercapacitor can reach 178 $F \cdot g^{-1}$, and it has good rate performance and cycle stability. Guoxing et al. prepared graphene/conductive polymer composite hollow fiber with the hydrothermal method. The combination of hollow structure and pseudocapacitance provided by conductive polymer greatly improved the capacity of the capacitor and provided a new idea for the improvement of supercapacitor capacitance [149].

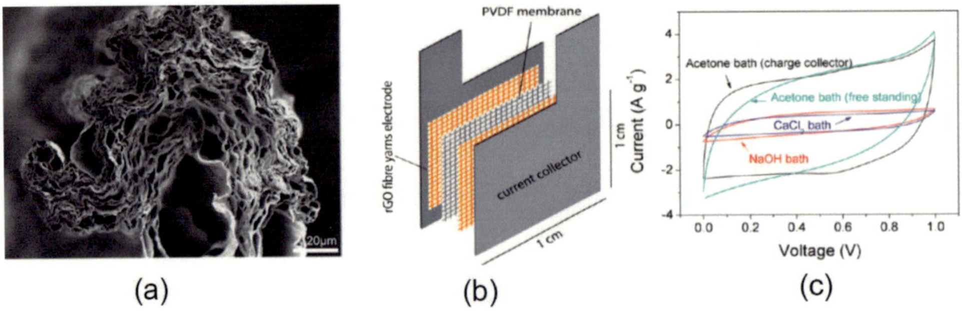

Figure 14.
Porous graphene fiber and its supercapacitor. (a) SEM image of porous fibers. (b) Schematic illustration of the structure of supercapacitor. (c) CV curves of graphene fibers prepared in different coagulation baths.

4.2.3 Energy conversion

4.2.3.1 Actuator

Actuators are a kind of stimuli-sensitive device that can respond to external stimuli, such as humidity, temperature, and electrical changes, and transfer the stimulus into deformation or motion [126, 127]. Due to quantum mechanics and electrostatic double-layer effect, graphene may cause space warping or plane expansion under the charge injection. In addition, the intercalation or removal of ions or molecules in graphene products under external stimulation will also lead to the bending, twisting, and even reversible change of the interlayer spacing. In this way, the type and degree of deformation can be controlled by the composition and surface chemical state of graphene [150, 151].

Jia et al. showed an electrochemical fiber driver with high driving activity and durability based on GF/polypyrrole (GF/PPY) double-layer structure, as shown in **Figure 15**. Because of the asymmetry of the structure, GF/PPY fiber shows reversible bending deformation under the condition on positive and negative charges. As shown in **Figure 15**, when a positive voltage is applied to GF/PPY fiber, graphene will shrink and expand due to anion discharged from PPY, and the fiber will bend to the left. When a negative voltage is applied, GF/PPY fiber can bend to the right [152].

Compared with rGO, GO has more oxygen functional groups, so it is more sensitive to water. Based on this principle, Huhu et al. fabricated an asymmetric rGO/GO fiber by region-selective laser reduction along the GO fiber. When exposed to humid air, the rGO/GO fiber can bend to the rGO side and then return to its original state after air moisture dispersion. After that, they made a twisted GO fiber by rotating the GO hydrogel fibers in the direction of rotation. The spiral geometry inside them was the main reason for the reversible rotation in the moist air.

4.2.3.2 Solar cell

Wearable solar cells can supply power to flexible smart devices at any time, while GBFs can be used as electrode materials to achieve this new function. Peng et al.

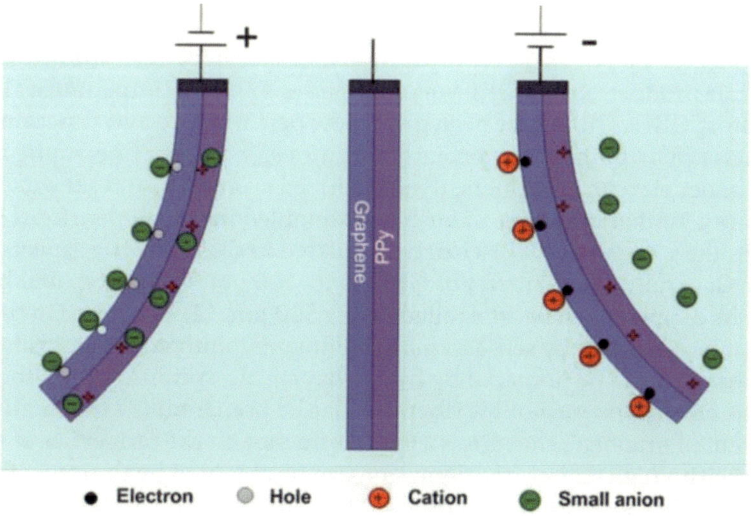

Figure 15.
Schematic illustration of the expansion-contraction mechanisms of the GF/PPY bilayer structure. Charges in each electrode are completely balanced by ions from the electrolyte.

obtained GBFs by wet spinning and then made its surface loaded with Pt metal particles by electrodeposition to the obtained counter electrode. The titanium wire with titanium dioxide microtubules on the surface is used as the working electrode; the dye-sensitized solar cell (DSSC) has an energy conversion efficiency of 8.45%, which is much higher than other linear photovoltaic devices. The continuous collection of energy can be realized by putting linear solar cells into conventional clothes [153]. The high surface properties and good electrical and electrochemical properties of graphene are the important reasons to improve the performance of fiber DSSC.

5. Conclusions and perspective

This chapter mainly summarizes the main preparation methods, properties, and application fields of CNT, CF, and GBF materials. Among them, CNTs have unique one-dimensional nanostructures and excellent mechanical, electrical, and optical properties. Through various methods of modification, researchers continue to prepare CNT composite nanomaterials with excellent performance, which has a good application prospect. Starting from the needs of the application field, it is the trend to study the carbon nanotube composite materials in the future to expect to obtain the high-efficiency structure which is corresponding to the application performance. Although some progress has been made in the preparation and properties of carbon nanotube composites, the mechanism of improving the properties of composites and the dispersion of carbon nanotubes still need to be explored.

CF is a new type of fiber material with high strength and high modulus, which contains more than 95% carbon. Its quality is lighter than that of aluminum, but its strength is higher than that of steel, and it has the characteristics of corrosion resistance and high-temperature resistance. It is an important material in the military industry and civil use. With the rapid development of CF composite and the continuous improvement of molding technology, its application scope is expanding day by day, and it shows good application potential in many fields. However, the physical and chemical properties of CF composites are complex, so it is necessary to study the basic theories of physical and chemical properties, mechanics, and heat, so as to improve the performance of CF composites.

GBFs have achieved great success in functional application, and it is far more amazing than CF. So far, various preparation methods have been studied and used in large-scale production of GBFs, which provides a positive impetus for the future application of GBFs. GBFs have been given new performance and function and provide new opportunities for various applications, including fiber-optic actuators, batteries, super electric containers, dye-sensitized solar cells, and sensors.

GBFs are a kind of graphene nanosheet assembled in one-dimensional space. At present, the structure of GBFs can be regulated in the following aspects: (1) Diameter. Generally, the diameter of GBFs is 10–100 μm. If it is prepared by electrospinning, its diameter can be controlled below 500 μm. (2) Porosity. On the one hand, it can be prepared by self-assembly, rolling, graphitization, and sintering; on the other hand, it can be prepared by freeze-drying, air spinning, and other methods. In addition, graphene hollow fiber can also be prepared. (3) Orientation. The arrangement of graphene sheets has a great influence on the properties of GBFs. The GBFs with a high degree of orientation can be obtained by the stress field orientation effect in the wet spinning process, the self-assembly in the electrochemical deposition process, and the second phase auxiliary orientation effect in the composite fiber. (4) Section morphology. It is difficult to maintain the circular cross section of the fiber, which is generally irregular. At present, the conventional method is to adjust the fiber cross-sectional shape by adjusting the spinneret hole shape, but the research progress is slow.

In order to meet the needs of different applications, graphene composite fibers appear. The additive materials include metal, inorganic, and polymer materials, such as silver nanowires, silicon nanoparticles, molybdenum disulfide nanoparticles, polypyrrole nanoparticles, etc. Basically, any nanomaterial can be added to GBFs to get graphene composite fiber. But one of the key problems is to control the structure of the composite fiber. The main control factor is the morphology of the second phase and its distribution in the fiber. For GBFs and its composite fiber, the main problems are as follows: (1) Compared with the graphene nanoflakes, the properties of GBFs are greatly cracked. (2) GBFs are composed of layers, which are very different from the chain structure of the traditional chemical fiber, so its flexibility is poor. (3) It is difficult to realize continuous production. Even with the most suitable wet spinning method for continuous production, its continuous production is very difficult, and the yield is very low.

Although GBFs are faced with many problems, remarkable achievements have been made. Compared with CF, GBFs have the characteristics of high strength, high modulus, conductivity, and certain flexibility, which have developed into a new type of high-performance fiber. On the other hand, graphene composite fiber is committed to develop into a new type of multifunctional intelligent fiber. This kind of fiber starts from modifying the traditional general-purpose fiber to improve some aspects of the performance of the general-purpose fiber and to develop new kinds of fiber, such as graphene/nano titanium oxide composite fiber. It can also develop new fiber performance and functions, such as energy storage, and finally realize multiple functions such as perception, judgment, correspondence, information transmission, etc. on the fiber and become a new type of intelligent material. Therefore, GBFs and its composite fiber will be widely used in aerospace, energy sensing, intelligent life, and other fields in the future.

Acknowledgements

The authors thank the support of Stevens startup fund.

Conflict of interest

The authors declare no conflict of interest.

Author details

Yan Xu and Xian Zhang*
Stevens Institute of Technology, Hoboken, NJ, Untied States

*Address all correspondence to: xian.zhang@stevens.edu

References

[1] Zhang X. Characterization of layer number of two-dimensional transition metal diselenide semiconducting devices using Si-peak analysis. Advances in Materials Science and Engineering. 2019;**2019**:7865698

[2] Li Y, Ye F, Xu J, Zhang W, Feng P, Zhang X. Gate-tuned temperature in a hexagonal boron nitride-encapsulated 2D semiconductor devices. IEEE Transactions on Electronic Devices. 2018;**65**(10):4068-4072

[3] Li Y, Chernikov A, Zhang X, Rigosi A, Hill HM, van der Zande AM, et al. Measurement of the optical dielectric function of monolayer transition-metal dichalcogenides: MoS_2, $MoSe_2$, WS_2, WSe_2. Physical Review B. 2014;**90**(20):205422

[4] Zhang X, Sun D, Li Y, Lee G-H, Cui X, Chenet D, et al. Measurement of lateral and interfacial thermal conductivity of single- and bilayer MoS_2 and $MoSe_2$ using optothermal Raman technique. ACS Applied Materials & Interfaces. 2015;7(46):25923-25929

[5] Zhang X, Chenet D, Kim B, Yu J, Tang J, Nuckolls C, et al. Fabrication of hundreds of field effect transistors on a single carbon nanotube for basic studies and molecular devices. Journal of Vacuum Science and Technology B. 2013;**31**(6):06FI01

[6] Yin D, Dun C, Gao X, Liu Y, Zhang X, Carroll D, et al. Controllable colloidal synthesis of Tin(II) chalcogenide nanocrystals and their solution-processed flexible thermoelectric thin films. Small. 2018;**14**(33):1801949

[7] Wu S, Wang L, Lai Y, Shan W-Y, Aivazian G, Zhang X, et al. Multiple hot-carrier collection in photo-excited graphene Moiré superlattices. Science Advances. 2016;**2**(5):e1600002

[8] De Vellis A, Gritsenko D, Lin Y, Wu Z, Zhang X, Pan Y, et al. Drastic sensing enhancement using acoustic bubbles for surface-based microfluidic sensors. Sensors and Actuators B: Chemical. 2017;**243**:298-302

[9] Gogotsi Y, Presser V. Carbon Nanomaterials. Florida, United States: CRC Press; 2013

[10] De Volder MF, Tawfick SH, Baughman RH, Hart AJ. Carbon nanotubes: Present and future commercial applications. Science. 2013;**339**(6119):535-539. DOI: 10.1126/science.1222453

[11] Gojny F, Wichmann M, Köpke U, Fiedler B, Schulte K. Carbon nanotube-reinforced epoxy-composites: Enhanced stiffness and fracture toughness at low nanotube content. Composites Science and Technology. 2014;**64**(15):2363-2371. DOI: 10.1016/j.compscitech.2004.04.002

[12] Hiremath N, Mays J, Bhat G. Recent developments in carbon fibers and carbon nanotube-based fibers: A review. Polymer Reviews. 2017;**57**(2):339-368. DOI: 10.1080/15583724.2016.1169546

[13] Gong K, Pan Z, Korayem AH, Qiu L, Li D, Collins F, et al. Reinforcing effects of graphene oxide on portland cement paste. Journal of Materials in Civil Engineering. 2014;**27**(2):A4014010. DOI: 10.1061/(ASCE)MT.1943-5533.0001125

[14] Lu C, Lu Z, Li Z, Leung CK. Effect of graphene oxide on the mechanical behavior of strain hardening cementitious composites. Construction and Building Materials. 2016;**120**:457-464. DOI: 10.1016/j.conbuildmat.2016.05.122

[15] Zhu Y, Murali S, Cai W, Li X, Suk JW, Potts JR, et al. Graphene and

graphene oxide: Synthesis, properties, and applications. Advanced Materials. 2010;**22**(35):3906-3924. DOI: 10.1002/adma.201001068

[16] Chatterjee S, Nüesch F, Chu BT. Comparing carbon nanotubes and graphene nanoplatelets as reinforcements in polyamide 12 composites. Nanotechnology. 2011;**22**(27):275714. DOI: 10.1088/0957-4484/22/27/275714

[17] Lingqi C, Xiaowei Y, Xiaomeng F, Meng C, Xiaokang M, Laifei C, et al. Mechanicalandelectromagneticshielding properties of carbon fiber reinforced silicon carbide matrix composites. Carbon. 2015;**95**:10-19. DOI: 10.1016/j.carbon.2015.08.011

[18] Rezaei F, Ynus R, Ibrahim NA. Effect of fiber length on thermomechanical properties of short carbon fiber reinforced polypropylene composites. Materials & Design. 2009;**30**(2):260-263. DOI: 10.1016/j.matdes.2008.05.005

[19] Waseem K, Rahul S, Parveen S. Carbon nanotube-based polymer composites: Synthesis, properties and applications. Carbon Nanotubes. 2016. DOI: 10.5772/62497

[20] Weng WZ, He SS, Song HY, Li XQ, Cao LH, Hu YJ, et al. Aligned carbon nanotubes reduce hypertrophic scar via regulating cell behavior. ACS Nano. 2018;**12**:7601-7612. DOI: 10.1021/acsnano.7b07439

[21] He SS, Zhang YY, Qiu LB, Zhang LS, Xie Y, Pan J, et al. Chemical-to-electricity carbon: Water device. Advanced Materials. 2018;**30**(18):1707635. DOI: 10.1002/adma.201707635

[22] Thostenson ET, Li WZ, Wang DZ. Carbon nano-tube/carbon fiber hybrid multiscale composites. Journal of Applied Physics. 2002;**91**(9):6034-6037. DOI: 10.1063/1.1466880

[23] Davis VA, Parra-Vasquez ANG, Green MJ, Rai PK, Behabtu N, Prieto V, et al. True solutions of single-walled carbon nanotubes for assembly into macroscopic materials. Nature nanotechnology. 2009;**4**:830-834. DOI: 10.1038/NNANO.2009.302

[24] Evans WJ, Hu L, Keblinski P. Thermal conductivity of graphene ribbons from equilibrium molecular dynamics: Effect of ribbon width, edge roughness, and hydrogen termination. Applied Physics Letters. 2010;**96**:203112. DOI: 10.1063/1.3435465

[25] Dalton AB, Collins S, Munoz E, Razal JM, Ebron VH, Ferraris JP, et al. Super-tough carbon-nanotube fibres-these extraordinary composite fibres can be woven into electronic textiles. Nature. 2003;**423**:703-703. DOI: 10.1039/B312092A

[26] Muhammad M, Ali R, Javed I. Production of carbon nanotubes by different routes-a review. Scientific Research. 2011;**1**:29-34. DOI: 10.4236/jeas.2011.11004

[27] Bin W, Yanfeng M, Yingpeng W, Na L, Yi H, Yongsheng C. Direct and large scale electric arc discharge synthesis of boron and nitrogen doped single-walled carbon nanotubes and their electronic properties. Carbon. 2009;**47**(8):2112-2115. DOI: 10.1016/j.carbon.2009.02.027

[28] Naiqin Z, Chunnian H, Zhaoyang J, Jiajun L, Yongdan L. Fabrication and growth mechanism of carbon nanotubes by catalytic chemical vapor deposition. Materials Letters. 2006;**60**(2):159-163. DOI: 10.1016/j.matlet.2005.08.009

[29] Qingwen L, Hao Y, Jin Z, Zhongfan L. Effect of hydrocarbons precursors on the formation of carbon nanotubes in chemical vapor deposition. Carbon. 2004;**42**(4):829-835. DOI: 10.1016/j.carbon.2004.01.070

[30] Yibo Y, Jianwei M, Zhihong Y, Fangxing X, Hongbin Y, Yanhui Y. Carbon nanotube catalysts: Recent advances in synthesis, characterization and applications. Chemical Society Reviews. 2015;**44**(10):3295-3346. DOI: 10.1039/c4cs00492b

[31] Andrea S, Caterina P, Anita C, Girolamo G, Danilo V, Janos BN. Synthesis methods of carbon nanotubes and related materials. Materials. 2010;**3**(5):3092-3140. DOI: 10.3390/ma3053092

[32] Somu C, Karthi A, Sanjay S, Karthikeyan R, Dinesh S, Ganesh N. Synthesis of various forms of carbon nanotubes by arc discharge methods—Comprehensive review. International Research Journal of Engineering and Technology. 2017;**4**(1):344-354

[33] Kerdcharoen T, Wongchoosuk C. Carbon nanotube and metal oxide hybrid materials for gas sensing. Semiconductor Gas Sensors, DOI. 2013:386-407. DOI: 10.1533/9780857098665.3.386

[34] Deepalekshmi P, Kishor KS, Yves G, Qipeng G, Sabu T. Carbon nanotube based elastomer composites—An approach towards multifunctional materials. Journal of Materials Chemistry C. 2014;**40**(2):8446-8485. DOI: 10.1039/C4TC01037J

[35] Li YL, Kinloch IA, Windle AH. Direct spinning of carbon nanotube fibers from chemical vapor deposition synthesis. Science. 2004;**304**(5668):276-278. DOI: 10.1126/science.1094982

[36] Mubarak NM, Abdullah EC, Jayakumar NS, Sahu JN. An overview on methods for the production of carbon nanotubes. Journal of Industrial and Engineering Chemistry. 2014;**20**(4):1186-1197. DOI: 10.1016/j.jiec.2013.09.001

[37] Muhammad OM, Sylvain H, Khalid L. Carbon nanofiber based buckypaper used as a thermal interface material. Carbon. 2011;**49**(12):3820-3828. DOI: 10.1016/j.carbon.2011.05.015

[38] Demczyk BG, Wang YM, Cumings J, Hetman M, Han W, Zettl A, et al. Direct mechanical measurement of the tensile strength and elastic modulus of multiwalled carbon nanotubes. Materials Science and Engineering: A. 2002;**334**(1-2):173-178. DOI: 10.1016/S0921-5093(01)01807-X

[39] Coleman JN, Khan U, Gunko YK. Mechanical reinforcement of polymers using carbon nanotubes. Advanced Materials. 2006;**18**(6):689-706. DOI: 10.1002/adma.200501851

[40] Dresselhaus MS, Dresselhaus G, Charlier JC, Hernandez E. Electronic, thermal and mechanical properties of carbon nanotubes. The Royal Society. 2004;**362**(1823):2065-2098. DOI: 10.1098/rsta.2004.1430

[41] Florian HG, Malte HG, Wichmann BF, Karl S. Influence of different carbon nanotubes on the mechanical properties of epoxy matrix composites—A comparative study. Composites Science and Technology. 2005;**65**(15-16):2300-2313. DOI: 10.1016/j.compscitech.2005.04.021

[42] Aliev AE, Lima MH, Silverman EM, Baughman RH. Thermal conductivity of multi-walled carbon nanotube sheets: Radiation losses and quenching of phonon modes. Nanotechnology. 2009;**21**(3):035709. DOI: 10.1088/0957-4484/21/3/035709

[43] Lindsay L, Broido DA, Natalio M. Diameter dependence of carbon nanotube thermal conductivity and extension to the graphene limit. Physical Review B. 2010;**82**:161402. DOI: 10.1103/PhysRevB.82.161402

[44] Hone J, Batlogg B, Benes Z, Johnson AT, Fischer JE. Quantized phonon spectrum of single-wall

carbon nanotubes. Science. 2000;**289**(5485):1730-1733. DOI: 10.1126/science.289.5485.1730

[45] Wei F, Mengmeng Q, Yiyu F. Toward highly thermally conductive all-carbon composites: Structure control. Carbon. 2016;**109**:575-597. DOI: 10.1016/j.carbon.2016.08.059

[46] Chang L, Huiming C. Carbon nanotubes: Controlled growth and application. Materials Today. 2013;**16**(1-2):19-28. DOI: 10.1016/j.mattod.2013.01.019

[47] Key HA, Won SK, Yong SP, Jeongmi M, Dong JB. Electrochemical properties of high-power super-capacitors using single-walled carbon nanotube electrodes. Advanced Functional Materials. 2001;**11**(5):387-391. DOI: 10.1002/1616-3028(200110)11:5<387::AID-ADFM387>3.0.CO;2-G

[48] Bin X, Feng W, Fang W, Shi C, Gaoping C, Yusheng Y. Single-walled carbon nanotubes as electrode materials for supercapacitors. Chinese Journal of Chemistry. 2006;**24**(11):1505-1508. DOI: 10.1002/cjoc.200690284

[49] Chichang H, Jenhong S, Tenchin W. Modification of multi-walled carbon nanotubes for electric double-layer capacitors: Tube opening and surface functionalization. Journal of Physics and Chemistry of Solids. 2007;**68**(12):2353-2362. DOI: 10.1016/j.jpcs.2007.07.002

[50] Minjung J, Euigyung J, Youngseak L. The surface chemical properties of multi-walled carbon nanotubes modified by thermal fluorination for electric double-layer capacitor. Applied Surface Science. 2015;**347**:250-257. DOI: 10.1016/j.apsusc.2015.04.038

[51] Guoping W, Lei Z, Jiujun Z. A review of electrode materials for electrochemical supercapacitors. Chemical Society Reviews. 2012;**41**(2):797-828. DOI: 10.1039/c1cs15060j

[52] Frank E, Steudle LM, Ingildeev D, Spörl JM, Buchmeiser MR. Carbon fibers: Precursor systems, processing, structure, and properties. Angewandte Chemie International Edition. 2014;**53**(21):5262-5298. DOI: 10.1002/anie.201306129

[53] Yun Z, Zhaojun L, Huiqi W, Jingli S, Jincai Z. Microstructure and thermal/mechanical properties of short carbon fiber-reinforced natural graphite flake composites with mesophase pitch as the binder. Carbon. 2013;**53**:313-320. DOI: 10.1016/j.carbon.2012.11.013

[54] He F. Carbon Fiber and Application Technology. Shanghai, China: Chemical Industry Press; 2004

[55] Soutis C. Carbon fiber reinforced plastics in aircraft construction. Materials Science and Engineering: A. 2005;**412**(1-2):171-176. DOI: 10.1016/j.msea.2005.08.064

[56] Krenkel W. Carbon fiber reinforced CMC for high-performance structures. International Journal of Applied Ceramic Technology. 2004;**1**(2):188-200. DOI: 10.1111/j.1744-7402.2004.tb00169.x

[57] Erik F, Frank H, Michael RB. Carbon fibers: Precursors, manufacturing, and properties. Macromolecular Materials and Engineering. 2012;**297**(6):493-501. DOI: 10.1002/mame.201100406

[58] Zhaokun M, Jingli S, Yan S. Carbon with high thermal conductivity, prepared from ribbon-shaped mesosphase pitch-based fibers. Carbon. 2006;**44**(7):1298-1301. DOI: 10.1016/j.carbon.2006.01.015

[59] Van Hattum FW, Bernardo CA, Finegan JC, Tibbetts GG, Alig RL,

Lake ML. A study of the thermo-mechanical properties of carbon fiber-polypropylene composites. Polymer Composites. 2004;**20**(5):683-688. DOI: 10.1002/pc.10391

[60] Basu S, Bhattacharyya P. Recent developments on graphene and graphene oxide based solid state gas sensors. Sensors and Actuators B: Chemical. 2012;**173**:1-21. DOI: 10.1016/j.snb.2012.07.092

[61] Noveselov KS, Geim AK, Morozov SV, Jiang D, Zhang Y. Electric field effect in atomically thin carbon films. Science. 2004;**306**(5696):666-669. DOI: 10.1126/science.1102896

[62] Wongbong C, Indranil L, Raghunandan S, Yong SK. Synthesis of Graphene and its applications: A review. Critical Reviews in Solid State and Materials Sciences. 2010;**35**:1. DOI: 10.1080/10408430903505036

[63] Fancheng M, Weibang L, Qingwen L, Joonhyung B, Youngseok O, Tsuwei C. Graphene-based fibers: A review. Advanced Materials. 2015;**27**:5113-5131. DOI: 10.1002/adma.201501126

[64] Alexander AB, Suchismita G, Wenzhong B, Irene C. Superior thermal conductivity of single-layer graphene. Nano Letters. 2008;**8**(3):902-907. DOI: 10.1021/nl0731872

[65] Lee CG, Wei XD, Kysar JW, Hong J. Measurement of the elastic properties and intrinsic strength of monolayer graphene. Science. 2008;**321**(5887):385-388. DOI: 10.1126/science.1157996

[66] Novoselov KS, Geim AK, Morozov SV, Jiang D, Katsnelson MI, Grigorieva IV, et al. Two-dimensional gas of massless dirac fermions in graphene. Nature. 2005;**438**:197-200

[67] Zhang YB, Tan YW, Stormer HL, Kim P. Experimental observation of the quantum Hall effect and Berry's phase in graphene. Nature. 2005;**438**:201-204

[68] Bolotin KI, Sikes KJ, Jiang Z, Klima M, Fudenberg G, Hone J, et al. Ultrahigh electron mobility in suspended graphene. Solid State Communications. 2008;**146**:351-355. DOI: 10.1016/j.ssc.2008.02.024

[69] Morozov SV, Novoselov KS, Katsnelson MI, Schedin F, Elias DC, Jaszczak JA, et al. Giant intrinsic carrier mobilities in graphene and its bilayer. Physical Review Letters. 2008;**100**:016602. DOI: 10.1103/PhysRevLett.100.016602

[70] Chen HJ, Jang C, Xiao SD, Ishigami M, Fuhrer MS. Intrinsic and extrinsic performance limits of graphene devices on SiO_2. Nature Nanotechnology. 2008;**3**:206-209. DOI: 10.1038/nnano.2008.58

[71] Si YC, Samulski ET. Synthesis of water soluble graphene. International Nano Letters. 2008;**8**:1679-1682. DOI: 10.1021/nl080604h

[72] Pei S, Zhao J, Du J, Ren W, Cheng HM. Direct reduction of graphene oxide films into highly conductive and flexible graphene films by hydrohalic acids. Carbon. 2010;**48**:4466-4474. DOI: 10.1016/j.carbon.2010.08.006

[73] Moon K, Lee J, Ruoff RS, Lee H. Reduced graphene oxide by chemical graphitization. Nature Communications. 2010;**1**:73-78. DOI: 10.1038/ncomms1067

[74] Becerril HA, Mao J, Liu Z, Stoltenberg RM, Bao Z, Chen YS. Evaluation of solution-processed reduced graphene oxide films as transparent conductors. ACS Nano. 2008;**2**:463-470. DOI: 10.1021/nn700375n

[75] McAllister MJ, Li J, Adamson DH, Schniepp HC, Abdala AA, Liu J, et al.

Single sheet functionalized graphene by oxidation and thermal expansion of graphite. Chemistry of Materials. 2007;**19**:4396-4404. DOI: 10.1021/cm0630800

[76] Zhu YW, Stoller MD, Cai WW, Velamakanni A, Piner RD, Chen D, et al. Exfoliation of graphite oxide in propylene carbonate and thermal reduction of the resulting graphene oxide platelets. ACS Nano. 2010;**4**:1227-1233. DOI: 10.1021/nn901689k

[77] Wang ZJ, Zhou XZ, Zhang J, Boey F, Zhang H. Direct electrochemical reduction of single-layer graphene oxide and subsequent functionalization with glucose oxidase. The Journal of Chemical Physics. 2009;**113**:14071-14075. DOI: 10.1021/jp906348x

[78] Haiyun S, Zhen X, Chao G. Multifunctional, ultra-flyweight. Synergistically Assembled Carbon Aerogels. Advanced Materials. 2013;**25**(18):2554. DOI: 10.1002/adma.201204576

[79] Stefania N, Daniel C, Luisa F, Maria CG, Francisco DM. Three dimensional macroporous architectures and aerogels built of carbon nanotubes and/or graphene: Synthesis and applications. Chemical Society Reviews. 2013;**43**(2):794-830. DOI: 10.1039/C2CS35353A

[80] Juergen B, Michael S, Matthew S, Marcus AW, Monika MB, Klint AR, et al. Advanced carbon aerogels for energy applications. Energy & Environmental Science. 2011;**4**:656-667. DOI: 10.1039/C0EE00627K

[81] Zakharchenko KV, Annalisa F, Los JH, Katsnelson MI. Melting of graphene: From two to one dimension. Journal of Physics: Condensed Matter. 2011;**23**(20):202202. DOI: 10.1088/0953-8984/23/20/202202

[82] Seong GK, David T. Melting the fullerenes: A molecular dynamics study. Physical Review Letters. 1994;**72**(15):2418. DOI: 10.1103/PhysRevLett.72.2418

[83] Liu YJ, Liang H, Xu Z, Xi JB, Chen GF, Gao WW, et al. Superconducting continuous graphene fibers via calcium intercalation. ACS Nano. 2017;**11**:4301-4306. DOI: 10.1021/acsnano.7b01491

[84] Meng J, Nie WQ, Zhang K, Xu FJ, Ding X, Wang SR, et al. Enhancing electrochemical performance of graphene fiber-based supercapacitors by plasma treatment. ACS Applied Materials & Interfaces. 2018;**10**:13652-13659. DOI: 10.1021/acsami.8b04438

[85] Choi SJ, Yu HY, Jang JS, Kim MH, Kim SJ, Jeong HS, et al. Nitrogen-doped single graphene fiber with platinum water dissociation catalyst for wearable humidity sensor. Small. 2018;**14**:1703934. DOI: 10.1002/smll.201703934

[86] Zhen X, Chao G. Graphene chiral liquid crystals and macroscopic assembled fibres. Nature Communications. 2011;**2**:571. DOI: 10.1038/ncomms1583

[87] Zhen X, Haiyan S, Xiaoli Z, Chao G. Ultrastrong fibers assembled from giant graphene oxide sheets. Advanced Materials. 2012;**25**(2):188. DOI: 10.1002/adma.201203448

[88] Xiang C, Young CC, Wang X, Yan Z, Hwang CC, Cerioti G. Large flake graphene oxide fibers with unconventional 100% knot efficiency and highly aligned small flake graphene oxide fibers. Advanced Materials. 2013;**25**(33):4592-4597. DOI: 10.1002/adma.201301065

[89] Zhen X, Yingjun L, Xiaoli Z, Li P, Haiyan S, Yang X, et al. Ultrastiff and strong graphene fibers via full-scale synergetic defect engineering. Advanced Materials. 2016;**28**(30):6449-6456. DOI: 10.1002/adma.201506426

[90] Yingjun L, Zhen X, Jianmig Z, Peigang L, Chao G. Superb electrically conductive graphene fibers via doping strategy. Advanced Materials. 2016;**28**(36):7941-7947. DOI: 10.1002/adma.201602444

[91] Shaohua C, Wujun M, Yanhua C, Zhe W, Bin S, Lu W, et al. Scalable non-liquid-crystal spinning of locally aligned graphene fibers for high-performance wearable supercapacitors. Nano Energy. 2015;**15**:642-653. DOI: 10.1016/j.nanoen.2015.05.004

[92] Xiaoteng D, Jie B, Tong X, Changxia L, Huimin Z, Liangti Q. A novel nitrogen-doped graphene fiber microelectrode with ultrahigh sensitivity for the detection of dopamine. Electrochemistry Communications. 2016;**72**:122-125. DOI: 10.1016/j.elecom.2016.09.021

[93] Xiaoteng D, Yang Z, Chuangang H, Yue H, Zelin D, Nan C, et al. Spinning fabrication of graphene/polypyrrole composite fibers for all-solid-state, flexible fibriform supercapacitors. Journal of Materials Chemistry A. 2014;**2**(31):12355-12360. DOI: 10.1039/C4TA01230E

[94] Wujun M, Shaohua C, Shengyuan Y, Wenping C, Wei W, Yanhua C. Flexible all-solid-state asymmetric supercapacitor based on transition metal oxide nanorods/reduced graphene oxide hybrid fibers with high energy density. Carbon. 2017;**133**:151-158. DOI: 10.1016/j.carbon.2016.11.051

[95] Wujun M, Shaohua C, Shengyuan Y, Meifang Z. Hierarchically porous carbon black/graphene hybrid fibers for high performance flexible supercapacitors. RSC Advances. 2016;**6**(55):50112-50118. DOI: 10.1039/C6RA08799J

[96] Wujun M, Shaohua C, Shengyuan Y, Wenping C, Wei W, Meifang Z. Bottom-up fabrication of activated carbon fiber for all-solid-state supercapacitor with excellent electrochemical performance. ACS Applied Materials & Interfaces. 2016;**8**(23):14622-14627. DOI: 10.1021/acsami.6b04026

[97] Wujun M, Shaohua C, Shengyuan Y, Wenping C, Yanhua C, Yiwei G, et al. Hierarchical MnO_2 nanowire/graphene hybrid fibers with excellent electrochemical performance for flexible solid-state supercapacitors. Journal of Power Sources. 2016;**306**(29):481-488. DOI: 10.1016/j.jpowsour.2015.12.063

[98] Shaohua C, Wujun M, Hengxue X, Yanhua C, Shengyuan Y, Wei W, et al. Conductive, tough, hydrophilic poly(vinyl alcohol)/graphene hybrid fibers for wearable supercapacitors. Journal of Power Sources. 2016;**319**(1):271-280. DOI: 10.1016/j.jpowsour.2016.04.030

[99] Guoyin C, Tao C, Kai H, Wujun M, Mike T, Yanhua C, et al. Robust, hydrophilic graphene/cellulose nanocrystal fiber-based electrode with high capacitive performance and conductivity. Carbon. 2018;**127**:218-227. DOI: 10.1016/j.carbon.2017.11.012

[100] Mochen L, Xiaohong Z, Xiang W, Yue R, Jinliang Q. Ultrastrong graphene-based fibers with increased elongation. Nano Letters. 2016;**16**(10):6511-6515. DOI: 10.1021/acs.nanolett.6b03108

[101] Yang Z, Changcheng J, Chuangang H, Zhlin D, Jiangli X, Yuning M, et al. Large-scale spinning assembly of neat, morphology-defined, graphene based hollow fibers. ACS Nano. 2013;**7**(3):2406-2412. DOI: 10.1021/nn305674a

[102] Dan L, Marc BM, Scott G, Richard BK, Gordon GW. Processable aqueous dispersions of graphene nanosheets. Nature Nanotechnology. 2008;**3**(2):101-105. DOI: 10.1038/nnano.2007.451

[103] Xiang C, Behabtu N, Lin Y, Chae HG, Young CC, Genorio B, et al. Graphene nanoribbons as an advanced precursor for making carbon fiber. ACS Nano. 2013;**7**(2):1628-1637. DOI: 10.1021/nn305506s

[104] Jang EY, Kim WJ, Kim T, Kang TJ, Kim YH, Carretero GJ, et al. Fibers of reduced graphene oxide nanoribbons. Nanotechnology. 2012;**23**(23):235601. DOI: 10.1088/0957-4484/23/23/235601

[105] Dong Z, Jiang C, Cheng H, Zhao Y, Shi G, Jiang L, et al. Facile fabrication of light, flexible and multifunctional graphene fibers. Advanced Materials. 2012;**24**(10):1856-1861. DOI: 10.1002/adma.201200170

[106] Yu D, Goh K, Wang H, Wei L, Jiang W, Zhang Q, et al. Scalable synthesis of hierarchically structured carbon nanotube-graphene fibres for capacitive energy storage. Nature Nanotechnology. 2014;**9**(7):555-562. DOI: 10.1038/nnano.2014.93

[107] Qishi T, Zhen X, Yingjun L, Bo F, Li P, Jiabin X, et al. Dry spinning approach to continuous graphene fibers with high toughness. Nanoscale. 2017;**9**(34):12335-12342. DOI: 10.1039/C7NR03895J

[108] Muge A, Geunsik L, Cecilia M, Adam P, Robert MW, Manish C, et al. The role of oxygen during thermal reduction of graphene oxide studied by infrared absorption spectroscopy. The Journal of Physical Chemistry C. 2011;**115**(40):19761-19781. DOI: 10.1021/jp2052618

[109] Hae-kyung J, Yun PL, Mei HJ, Eun SK, Jung JB, Young HL. Thermal stability of graphite oxide. Chemical Physics Letters. 2009;**470**(4-6):255-258. DOI: 10.1016/j.cplett.2009.01.050

[110] Qing C, Yuning M, Chuangang H, Yang Z, Huibo S, Nan C, et al. MnO_2-modified hierarchical graphene fiber electrochemical supercapacitor. Journal of Power Sources. 2014;**247**:32-39. DOI: 10.1016/j.jpowsour.2013.08.045

[111] Jihao L, Jingye L, Linfan L, Ming Y, Hongjuan M, Bowu Z. Flexible graphene fibers prepared by chemical reduction-induced self-assembly. Journal of Materials Chemistry A. 2014;**2**(18):6359-6362. DOI: 10.1039/C4TA00431K

[112] Chuangang H, Yang Z, Huhu C, Yanhong W, Zelin D, Changcheng J, et al. Graphene microtubings: controlled fabrication and site-specific functionalization. Nano Letters. 2012;**12**(11):5879-5884. DOI: 10.1021/nl303243h

[113] Jing F, Zelin D, Meiling Q, Ruonong F, Liangti Q. Monolithic graphene fibers for solid-phase microextraction. Journal of Chromatography A. 2013;**1320**:27-32. DOI: 10.1016/j.chroma.2013.10.065

[114] Xiaosong H. Fabrication and properties of carbon fibers. Materials. 2009;**2**(4):2369-2403. DOI: 10.3390/ma2042369

[115] Shayan S, Mark SR, Andrew IM, Joselito MR. Towards the knittability of graphene oxide fibres. Scientific Reports. 2015;**5**:14946. DOI: 10.1038/srep14946

[116] Xinming L, Tianshuo Z, Kunlin W, Ying Y, Jinquan W, Feiyu K, et al. Directly drawing self-assembled, porous, and monolithic graphene fiber from chemical vapor deposition grown graphene film and its electrochemical properties. Langmuir. 2011;**27**(19):12164-12171. DOI: 10.1021/la202380g

[117] Seyed HA, Rouhollah J, Dorna E, Maryam S, Zahra G, Sima AY, et al. High-performance multifunctional graphene yarns: Toward wearable all-carbon energy storage textiles.

ACS Nano. 2014;**8**(3):2456-2466. DOI: 10.1021/nn406026z

[118] Jiali Y, Mei W, Ping X, Seung HC, Jonghwan S, Ke G, et al. Ultrahigh-rate wire-shaped supercapacitor based on graphene fiber. Carbon. 2017;**119**:332-338. DOI: 10.1016/j.carbon.2017.04.052

[119] Yunming J, Mei Z, Hongwei L, Jianming W, Fanglan G. Controllable synthesis and electrochemical performance of hierarchically structured graphene fibers. Materials Chemistry and Physics. 2017;**193**:35-41. DOI: 10.1016/j.matchemphys.2017.02.014

[120] Lizhi S, Tong W, Yuan L, Lili J, Liangti Q, Zhuangjun F. Ultra-high toughness all graphene fibers derived from synergetic effect of interconnected graphene ribbons and graphene sheets. Carbon. 2017;**120**:17-22. DOI: 10.1016/j.carbon.2017.05.033

[121] Junjie Y, Wei W, Yang Z, Xiaowen D, Yunxia L, Lijun Y, et al. Highly flexible and shape-persistent graphene microtube and its application in supercapacitor. Carbon. 2018;**126**:419-425. DOI: 10.1016/j.carbon.2017.10.045

[122] Lianlian C, Yu L, Yang Z, Nan C, Liangti Q. Graphene-based fibers for supercapacitor applications. Nanotechnology. 2015;**27**(3):032001. DOI: 10.1088/0957-4484/27/3/032001

[123] Fei Z, Yang Z, Huhu C, Liangti Q. A graphene fibriform responsor for sensing heat, humidity, and mechanical changes. Angewandte Chemie International Edition. 2015; **54**(49):14951-14955. DOI: 10.1002/anie.201508300

[124] Yanhong W, Ke B, Chuangang H, Zhipan Z, Nan C, Huimin Z, et al. Flexible and wearable graphene/polypyrrole fibers towards multi-functional actuator applications. Electrochemistry Communications. 2013;**35**:49-52. DOI: 10.1016/j.elecom.2013.07.044

[125] Chunfei H, Yuanyuan S, Xiying L, Xiaoyang H, Ying W, Xinchang W, et al. Helical graphene oxide fibers as a stretchable sensor and an electrocapillary sucker. Nanoscale. 2016;**8**(20):10659-10668. DOI: 10.1039/C6NR02111E

[126] Huhu C, Jia L, Yang Z, Chuangang H, Zhipan Z, Nan C, et al. Graphene fibers with predetermined deformation as moisture-triggered actuators and robots. Angewandte Chemie International Edition. 2013;**52**(40):10482-10486. DOI: 10.1002/anie.201304358

[127] Huhu C, Yue H, Fei Z, Zelin D, Yanhong W, Nan C, et al. Moisture-activated torsional graphene-fiber motor. Advanced Materials;**26**(18):2909-2913. DOI: 10.1002/adma.201305708

[128] Fei L, Shuyan S, Dongfeng X, Hongjie Z. Folded structured graphene paper for high performance electrode materials. Advanced Materials. 2012;**24**(8):1089-1094. DOI: 10.1002/adma.201104691

[129] Wei A, Zhimin L, Jian J, Jianhui Z, Zhuzhu D, Zhanxi F, et al. Nitrogen and sulfur codoped graphene: Multifunctional electrode materials for high-performance li-ion batteries and oxygen reduction reaction. Advanced Materials. 2014;**26**(35):6186-6192. DOI: 10.1002/adma.201401427

[130] Jong GL, Youbin K, Ji YJ, Sungho C, Yongku K, Woong RY, et al. Fiber electrode by one-pot wet-spinning of graphene and manganese oxide nanowires for wearable lithium-ion batteries. Journal of Applied Electrochemistry. 2017;**47**(8):865-875. DOI: 10.1007/s10800-017-1085-y

[131] Minsu G, Seunghee K, Seungmin Y, Eunhee L, Sa HM, Soojin P, et al. Double locked silver-coated silicon nanoparticle/graphene core/shell fiber for high-performance lithium-ion battery anodes. Journal of Power Sources. 2015;**300**:351-357. DOI: 10.1016/j.jpowsour.2015.09.083

[132] Tatsumasa H, Yuanchuan Z, Junyu H, Zhiqiang W, Qingwen L, Zhigang Z, et al. Flexible lithium-ion fiber battery by the regular stacking of two-dimensional titanium oxide nanosheets hybridized with reduced graphene oxide. Nano Letters. 2017;**17**(6):3543-3549. DOI: 10.1021/acs.nanolett.7b00623

[133] Woon GC, Jian QH, Zheng LX, Xiangyin Q, Xiangyu W, Jang KK. Lithium–sulfur battery cable made from ultralight, flexible graphene/carbon nanotube/sulfur composite fibers. Advanced Functional Materials. 2016;**27**(4):1604815

[134] Biao Z, Feiyu K, Jean MT, Jang KK. Recent advances in electrospun carbon nanofibers and their application in electrochemical energy storage. Progress in Materials Science. 2016;**76**:319-380. DOI: 10.1016/j.pmatsci.2015.08.002

[135] Yong SK, Ghazal S, Yevgen Z, Jaehyuk L, Zhong L, Bharat P, et al. The critical contribution of unzipped graphene nanoribbons to scalable silicon–carbon fiber anodes in rechargeable Li-ion batteries. Nano Energy. 2015;**16**:446-457. DOI: 10.1016/j.nanoen.2015.07.017

[136] Xiaoyan W, Lian F, Decai G, Jian Z, Qingfeng Z, Bingan L. Core–Shell Ge@Graphene@TiO_2 nanofibers as a high-capacity and cycle-stable anode for lithium and sodium ion battery. Advanced Functional Materials. 2016;**26**(7):1104-1111. DOI: 10.1002/adfm.201504589

[137] Xiaoxin M, Guangmei H, Qing A, Lin Z, Pengchao S, Jinkui F, et al. A heart-coronary arteries structure of carbon nanofibers/graphene/silicon composite anode for high performance lithium ion batteries. Scientific Reports. 2017;**7**(1):9642. DOI: 10.1038/s41598-017-09658-4

[138] Jian Z, Guanhua Z, Xinzhi Y, Qinhong L, Bingan L, Zhi X. Graphene double protection strategy to improve the SnO_2 electrode performance anodes for lithium-ion batteries. Nano Energy. 2014;**3**:80-87. DOI: 10.1016/j.nanoen.2013.10.009

[139] Ji C, Chun L, Gaoquan S. Graphene materials for electrochemical capacitors. The Journal of Physical Chemistry Letters. 2013;**4**:1244-1253. DOI: 10.1021/jz400160k

[140] Liang H, Chun L, Gaoquan S. High-performance and flexible electrochemical capacitors based on graphene/polymer composite films. Journal of Materials Chemistry A. 2014;**2**:968-974. DOI: 10.1039/C3TA14511E

[141] Jing L, Xianke H, Linfan C, Nan C, Liangti Q. Preparation and supercapacitor performance of assembled graphene fiber and foam. Progress in Natural Science: Materials International. 2016;**3**:212-220. DOI: 10.1016/j.pnsc.2016.05.006

[142] Yue H, Huhu C, Fei Z, Nan C, Lan J, Zhihai F, et al. All-in-one graphene fiber supercapacitors. Nanoscale. 2014;**6**:6448-6451. DOI: 10.1039/C4NR01220H

[143] Yang Z, Qing H, Zhihua C, Lan J, Liangti Q. Integrated graphene systems by laser irradiation for advanced devices. Nano Today. 2017;**12**:14-30. DOI: 10.1016/j.nantod.2016.12.010

[144] Yunzhen C, Gaoyi H, Dongying F, Feifei L, Miaoyu L, Yanping L. Larger-scale fabrication of N-doped graphene-fiber mats used in high-performance

energy storage. Journal of Power Sources. 2014;**252**:113-121. DOI: 10.1016/j.jpowsour.2013.11.115

[145] Guan W, Pengfeng T, Xingjiang W, Lu P, Hengyang C, Cai FW, et al. High-performance wearable micro-supercapacitors based on microfluidic-directed nitrogen-doped graphene fiber electrodes. Advanced Functional Materials. 2017;**27**(36):1702493. DOI: 10.1002/adfm.201702493

[146] Yuning M, Yang Z, Chuangang H, Huhu C, Yue H, Zhipan Z, et al. All-graphene core-sheath microfibers for all-solid-state, stretchable fibriform supercapacitors and wearable electronic textiles. Advanced Materials. 2013;**25**(16):2326-2331. DOI: 10.1002/adma.201300132

[147] Bingjie W, Qingqing W, Hao S, Jing Z, Jing R, Yongfeng L, et al. An intercalated graphene/(molybdenum disulfide) hybrid fiber for capacitive energy storage. Journal of Materials Chemistry A. 2017;**5**(3):925-930. DOI: 10.1039/C6TA09360D

[148] Qiuyan Y, Zhen X, Bo F, Tieqi H, Shengying C, Hao C, et al. MXene/graphene hybrid fibers for high performance flexible supercapacitors. Journal of Materials Chemistry A. 2017;**5**(42):22113-22119. DOI: 10.1039/C7TA07999K

[149] Guoxing Q, Jianli C, Xiaodong L, Demao Y, Peining C, Xuli C, et al. A Fiber supercapacitor with high energy density based on hollow graphene/conducting polymer fiber electrode. Advanced Materials. 2016;**28**(19):3646-3652. DOI: 10.1002/adma.201600689

[150] Huhu C, Yaxin H, Gaoquan S, Lan J, Liangti Q. Graphene-based functional architectures: Sheets regulation and macrostructure construction toward actuators and power generators. Accounts of Chemical Research. 2017;**50**(7):1663-1671. DOI: 10.1021/acs.accounts.7b00131

[151] Luhua L, Jinghai L, Ying H, Yuewei Z, Wei C. Graphene-stabilized silver nanoparticle electrochemical electrode for actuator design. Advanced Materials. 2013;**25**(9):1270-1274. DOI: 10.1002/adma.201203655

[152] Jia L, Zhi W, Xuejun X, Huhu C, Yang Z, Liangti Q. A rationally-designed synergetic polypyrrole/graphene bilayer actuator. Journal of Materials Chemistry. 2012;**22**(9):4015-4020. DOI: 10.1039/C2JM15266E

[153] Zhibin Y, Hao S, Tao C, Longbin Q, Yongfeng L, Huisheng P. Photovoltaic wire derived from a graphene composite fiber achieving an 8.45% energy conversion efficiency. Angewandte Chemie International Edition. 2013;**52**(29):7545-7548. DOI: 10.1002/anie.201301776

Section 3

Polymer Nanocomposites

Chapter 6

Reinforce Fabricated Nano-Composite Matrixes for Modernization of S & T in New Millennium

Rajendra Sukhadeorao Dongre

Abstract

Rational fabrication of futuristic smart materials like polymer-derived nanocomposites/matrixes is ever desirable due to innate worth in advancement and growth of S&T. Certain nanocomposites are designed from metal-polymeric blends through altering varied parameters like shear stress, shape, size, rate, concentration and processing time, which are best used as fillers. Reinforced fabricated polymer nanocomposites possess exclusive physicochemical characteristics like non-Newtonian/constant viscosity-free stress, time-dependent mechanics, facile shear-skeletal revisions and viscoplastic course controls. Metal-derived nanocomposites/matrixes showed substantial inherent rheology being vulnerable for designing viable applicability in photovoltaics, catalysis, optics, drug delivery, smart material and energy storage. Bottom-up technique is used for self- and directed-assembly of polymer-based building blocks owing to robust fabricated and efficiently manipulated/targeted reinforced 1D, 2D or 3D nanostructures. This chapter reviews some contemporary advances in reconfiguration of rational designing of certain polymeric nanostructures/composites with current and futuristic developments. This overview highlights significance of assured reinforced matrixes in S&T besides disclosed fundamental principles involved in material designing/engineering of multifaceted nanomaterials. Assorted advanced developments are made to avail futuristic prospective of biopolymers, viz. chitin, chitosan, cellulose and lignin in order to offer unequivocally myriad applications in modernization of science and technology in new millennium.

Keywords: biopolymer, composite, chitosan, cellulose, nanocarbon, matrix, dendrimer

1. Introduction

The Latin term *nannus* meaning dwarfness is in fact adopted for the prefix *nano* [1]. Nanotechnology deals/controls the matter at both atomic and molecular developments, creating and modifying structural components in all possible dimension levels (1D, 2D, 3D etc.). Nanoparticles/nanomaterials showed extraordinary features due to inbuilt molecular/atom-by-atom accuracy that is deficient in conventional bulk counterparts/materials [2]. Nanotechnology manipulates structures/skeletons of matter at atom/molecular scale. "Nanotechnology" was coined by Norio Taniguchi in 1974, and in 1959, Richard Feynman gave the famous quote "There is Plenty of Room/space at the Bottom," which is trusted in assorted making of nano-scale machines. IBM Zurich researchers in 1980 invented tunneling microscope for material analysis at atomic/molecular dimension [1, 2]. Nano-composite matrix holds particles in one or more filler layers/sheets with superior surface:volume ratio at a magnitude of few nm units. Nano-dimensional alterations are found to reduce material size and shape without changing its native features and besides fairly reward novel features including alter elasticity, robust mechanical power, tunable heat and electric conductance or insulation, and impart particular reactivity that are absent in corresponding micro-/macro-scale dimensions. The interfacial-phase interactions that exist at nano-dimension scales are superior due to augmented intrinsic characteristics of the material, as multiphase combinations of constituents in fabricating nano-composite matrixes impart certain innovative qualities that are quite superior to residual participating constituents.

Assured nanostructures/skeletal matrixes derived via reconfiguration/reinforcement establish a myriad of functional importance in advancement of today's science and technology [2, 3]. Nanotechnology executes superior technological reconfigurations through strategic maneuvering of matter at an atom, molecule or supramolecular dimension at a magnitude less than 100 nm [1–3]. Reinforcement of atomic/molecular frameworks and material manipulation are performed at nanometer (10^{-9} m) scale via nanotechnology, which exactly manufactures micro-, meso- and macromaterials under its vast domain. In general, nanostructure matrixes exist in the form of amorphous, crystalline and polycrystalline states that embed variable size/shape including metallic, ceramics and polymers, besides offering single or multi-phase chemical compositions and designed orientations [2]. Nowadays, all such reinforced and reconfigured nanomatrixes own innate distinguishing scientific edicts in scientific and technological modernizations. The defensive features of meso-/micromaterial get enhanced at the nanoscale due to alteration of limiting features via augmenting physicochemical, biological, mechanical, electrical and electronic parameters. Reinforced/reconfigured nanomaterials own specially intended characteristics, viz. huge surface area, no/less surface defects and high surface/mass ratio that are best exploited in nearly all S&T achievements [4, 5].

Today, nanotechnology is expanded with a novel horizon, yet R&D in materials science is in a much infantile phase, though nanoscience capably upgraded standards of every domain including energy production/storage, information technology, pharmaceutics, metamaterial, nanomaterial, food, biotechnology and wastewater/water, biomedical, environment and instrument/device [1–4, 6]. Advance re-configurated designing yields diverse matrixes that are used in myriad applications like coatings, sunscreens, cosmetics, textiles, paints, cutting boards, socks, diodes, pacemaker, scaffold for hip/bone/ear joints and electrodes for H_2O splitting. Amid nanomaterials are nanocarbon, nanosilica and identified nanometals like copper, silver and gold, besides awarded nanometal oxides of iron, cerium, nickel, aluminum, titanium, zinc etc. along with a unique entity called quantum dots [3, 4, 6, 7]. All such fabricated composites/hybrids/matrixes being ubiquitous are tailored for endowing best vigor, stiffness and design practices, which have been trusted throughout the modernization of S&T.

2. Reinforced and designed composites/matrixes

Rationally designed/reinforced composites own specific strength and modulus over analogous materials like metallic alloys, steel and other metallic compositions [4, 6, 7]. Certain nonfiber matrixes are reinforced owing to varied highly anisotropic structural features that differ from isotropic polymers, metals and ceramics.

All such rationally fabricated reconfigured matrixes display extraordinary features like environment sturdiness, damage tolerance, non-erosive, sturdy thermo-oxidation, non-flammable and attenuated electric or heat conductance besides offering especial utility in proactive surface designing and possessing integrated networking. Such reinforced matrixes/composites offer many applications like water purifications, supercapacitors and green energy/power creation and anti-corrosive, antistatic, antiballistic and electroconductive devices.

The twenty-first century seeks special/smart materials with properties subjective to the remarkable varying conditions of their functioning. Such smart materials are reconfigured via advanced nanoscience innovations due for stunning and unique designing opportunities for advancement of science. Reinforced composites/matrixes have eventually modernized S&T via a myriad of applications including optoelectronics, biosensors, photodetectors, photocells, nanomaterials and plasmonics [1–4, 6–9]. Nanocomposite/matrix offers environmental-friendly prospective for all sectors including chemistry-, physics-, bioscience-, engineering and technology-based industries and businesses. Nanomaterials have explored, inspired and motivated interdisciplinary R&D and accordingly trusted scientific growth. Nanosize designing improved physicochemical properties of materials through exceptionally determined functionalities and paved new trends besides modern developments. Today, R&D comprehensive findings conferred assorted innovative nanocomposites of assorted materials including chitosan, cellulose, metals, polymers, clays, carbon and graphene owing to numerous applications [10]. Nano-technologically developed material matrixes behold systematic captivity at nano-scale due to smartly carried advanced reconfigurations or reinforcements in their skeletons. Reinforced polymeric matrix own rigid dispersed phase owing constituted particles with nanodimensions exists with long range flexible-rough linkages. Nanopolymer networks that are amorphous/semicrystalline own reasonably stacked and intercalated inter-phase morphology. Entropic morphological alterations are viable due to interactive component configuration [3].

Reconfigured composite contains incessant fortified pattern owing to reinforced physicochemical features that yield through two/more discrete constituting phases of mainly fibers, whiskers and metamaterials, polymers, metals and ceramics [1–4, 6–10]. In 1960, Richards Feynman's research in field of quantum computations really fascinated the entire scientific community toward nanotechnology; besides aiding to develop several material-based blends/composites/matrixes in order to furnish extensive applicability at that time, they still continued to explore various advanced nanostructures to be used in modernization of today's S&T. Sophisticated unique characters get inculcated in nanocomposites/matrixes imparting high modulus, specific strength, tailored performance, boosted resistivity in fatigue and corrosion besides embryonic reconfigured processing. Nanotechnology reinforcements can let mass diminution that is missing in usual counterstructures and contemporary materials, thus creating good substitutes. Supplementary complex functioning can be significantly enhanced by means of reinforcements that are unattainable in normal materials. Innovative structural strengthening can be attained via anisotropic or isotropic alterations at all altered phases in the constituted/reconfigured frameworks. These reinforced composites /matrixes offer special infrastructure developments in material designing and engineering like bridges, pipelines, transportation materials, automobiles, cermets, aircrafts and ships [3, 4, 6, 7].

3. Nanocarbon intercalated polymeric matrix

Assorted material components easily enter polymeric/biomolecule skeleton in synergistic pattern yielding nanobiocomposites and imparting advance structural features as an alternative to classic synthetic/natural polymers. Consequently, various inorganic materials including metal particles, carbon nanotubes, ceramics and clays are blended in biopolymers resulting in a diverse nanocomposite/hybrid like polymer-inorganic, metal-polymer, metal-ceramic and inorganic-organic phases [10]. All these rationally designed/reconfigured nanocomposites/hybrids/matrixes endow many applications, viz. biosensor, marker, biochip, optic, electric, electronic, photoconductors, biocompatible tissue engineered scaffolds/templates and drug release/filter. Monomeric inorganic/organic hosts/frameworks can be reinforced with many natural/bio- and synthetic polymers resulting in intercalated polymer networking composites [11]. Such matrixes are obtained via assorted techniques, viz. microwave, colloid interaction, suspended polymerization, solvent evaporation, electro-spinning, spray-drying, porous glass membrane spraying and emulsification. So, superior techniques are used for the development of desired characteristics like not expensive, competent, control/tunable shapes/sizes, porosity, density and surface area in reconfigured matrixes as devoid in counterparts.

Rationally reinforced polymeric nanocomposites hold host-guest intercalated morphological permutations and combinations of inorganic/organic frameworks like nanocarbon, metal, clay, montmorillonite, ceramic, poly-vinyl alcohol/chloride and zeolite [12]. Template or chosen material that holds native stupendous physicochemical characters is too vulnerable in the derived matrix. Reinforced composites/matrixes offer distinctive significance in electric, electronic gadgets, tissue engineering, packaging, coatings, biomedicals, nanodevice feedstock, photosensitivity, catalysts and antimicrobials and disinfectants besides physicochemical analysis. Various technologically reinforced 1D, 2D and 3D composites/matrixes own boosted intrinsic features and corresponding applicability domain as shown in **Figure 1**.

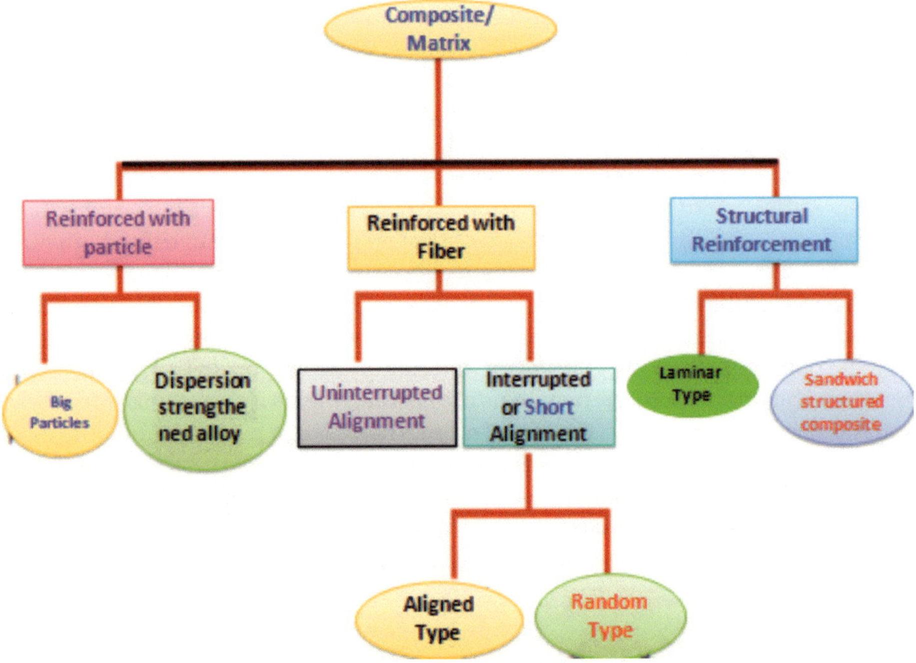

Figure 1.
Technologically reinforced various composite matrixes [1, 2].

4. New millennium advanced material matrixes in S&T

Today, smart materials have manifested assorted benefits and it is hard to envisage the modernized advancements without their contributions. Advanced materials endow a myriad of applicability in industries like chemical, mines, metallurgy, oil-gas extraction, refinery, power, and modern technology, viz. aerospace, IT, communication construction, transportation and genetic engineering. Smartly designed/reconfigured matrixes have to face few technically notable challenging domains being adept at power turbines and well robust aerojet engines etc. Certain smartly designed super-alloys are found to fulfill numerous such methodological challenges and demands that own practically efficient utilities in an industry besides R&D. Accordingly emerges prototype thrust in R&D of material and prevalent advanced nanotechnology assisted rational fabrications of smartly functional matrixes thus continued Richards Feynman initiated timeline advancement of nano-materials. Today, paradigm nano-technological developments have stimulated rational reconfiguration of materials and ultimately pave a path for designing classic, competing and preferred matrixes or composites for strengthening S&T in the new millennium.

This twenty-first century, invoke technological advancement in smartly designing and rational reconfigurations of nano-material matrixes to be developed via amalgamating incredible features of constituents in resultant composites (as meso/micro-porous materials like alloys, blends, ceramics, natural and synthetic polymers found to miss such designed features. More smart materials like 1D, 2D or 3D have to be architectured via reinforcement of two/more phases in vigor and firmly intercalated material framework as achieved in material engineering for sturdy, reinforced and robust output [3, 4, 6–12]. Augmented and perceived performance is practicable in such composites by means of particulate segregation due to tailoring of raw-skeletal elements. Advanced and sophisticated techniques aid in designing and reconfiguring assorted materials including natural and artificial origin. These reconfigured composites/matrixes are beneficial due to lesser density, superior directional mechanics, precisely enhanced tensile strength than steel/metals, elevated fatigue survival, adaptable tailoring/designing, facile machining and cost-effective synthesis. Directional arrangements of constituting matrix mutually control mechanical strength and functional properties of resultant composites. Parallel longitudinal atomic/molecular arrangements are obtained via solitary pathway and fully random configurations that are generally allied in the following sense: (a) Owing to irregular associations (b) Easy arbitrary/partial adjustments (c) Very much strengthening resultant composites due to small diameter, less surface flaws and facile suppleness over bulk materials (as seen in glass, aramid/kevlar and carbon fiber). Assorted 1D, 2D or 3D reconfigured material matrixes/composites are discussed in the following sections:

4.1 Fibrous composites

Certain fibers owing to greater length than diameter and (***l/d***) ratio imparting valuable shear pressure reassign reinforcement in arbitrary direction in their skeletons resulting in the most persuade fibrous composites as shown in **Figure 2**.

4.2 Carbon-reinforced composites

Carbon-reinforced composites have high reinforcement in their polymer matrix due to their innate tensile modulus and elevated strength at eminent temperatures, which are unaffected by water or other solvents, acids and bases. Carbon-reinforced composites display a variety of physicochemical and mechanical characters as

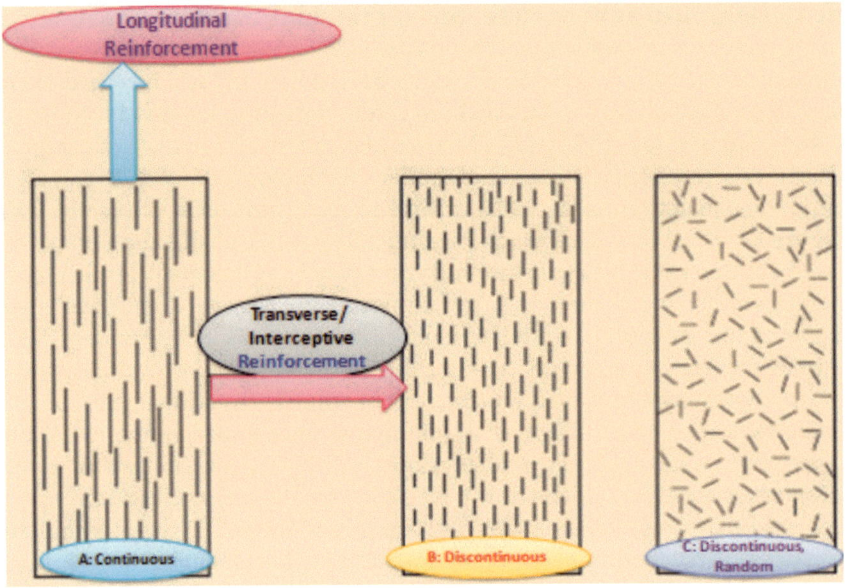

Figure 2.
Assorted orientations in fibrous reinforced composites [1, 2, 13].

exploited to get specific/rationally designed products like thermoplastic at quite a cheap cost. Such carbon-reinforced composites have huge strength:weight ratio and rigidity as needed in aerospace, transportation superstructures, automotive, engineered products, scaffolds and smart equipment. Certain fiber-reinforced plastic/composites are fabricated through polymeric matrix like organic/inorganic fibers, paper, wood and asbestos, which caters to the needs of aerospace, automotive, marine and constructions besides ballistic armor. American Chemist Leo Hendrik Baekeland in 1905 had replaced shellac-resin (yield from lac bug's excretion) with synthetic Bakelite polymer obtained via phenol-formaldehyde reaction at controlled pressure and temperature, and it was the world's first synthetic plastic that was fiber-reinforced [14]. In 1936, du-Pont obtained resin-composite through "*fiber-glass*" blending with plastic followed by modern *cyanamids* resin in 1942. Glass, carbon and aramid fibers are still used in making fiber-reinforced plastic/composites. Certain polymer-reinforced combinations are stated in **Table 1**.

4.3 Metal matrix composites

Metal-based matrixes/composites are obtained through ductile metal particulate fortification via continuous and discontinuous or whisker fiber molds. Reinforced composites appear as green/benign materials owing to special features like precise rigidity, nonflammability, high stability, abrasion/creep resistance and thermal/electrical conductivity besides sustainability at serviceable temperature and pressure than their counterparts [15]. But metal matrix composites are much more expensive than other reinforced composites, so they possess limited utility. Superalloys and alloys of metals are engaged in making such metal-based matrixes/composites [15, 16]. Continuous fiber moldings utilize assorted organic and inorganic fibers like carbon, silicon carbide, boron, aluminum oxide and certain refractory metals, while discontinuous path of reinforcements involves fibers of silicon carbide, aluminum oxide, silicon oxide and carbon [17]. Metal matrix composite yield via dispersion of reinforced fabric in metallic template and reinforced surfaces gets coating to avoid auto-oxidation. The template is made

Reinforced matrix/ composite	Constituting components	Superior features
Glass-based fiber	Epoxide, polyamide, polycarbonate, polyoxymethylene, polypropylene, vinyl ester, polybutylene terephthalate	Strengthen expansion coefficient, own great electric and heat resistance
Wood-based fiber	Polyethylene, polylactic acid, polypropylene, acrylonitrile butadiene styrene, high-density polyethylene	Flexural potency, high tensile modulus and strength
Carbon- and aramid-based fiber	polyepoxide, polyester, vinyl ester, polyacrylate	Suppleness, high tensile strength and solidity, electrical strengthen
Inorganic materials	Semicrystalline thermoplastics, UP	Isotropic contraction, graze, high solidity potential strength
Chitosan	Metals, nonmetals, synthetic and natural polymers	Flexible strength, alternative surface characters
Microsphere	Glass micro-/mesospheres	Less weight, solid fillers

Table 1.
Various reinforced composites with superior applicable features [13].

up of monolithic material wherein the reinforced fiber gets embedded through continuous moldings. Highly structured metal matrix/composite is derived through aluminum, magnesium or titanium metallic supports for reinforcement with fibers. The reinforced fiber can be embedded into metallic skeletons achieving either constant or irregular structural tasking, which can modify certain physico-chemical features, viz. wear/shear resistance, friction coefficient, and thermal and electrical conductivity [18].

Metal composite is reinforced through continuous or discontinuous mode of fabrications, viz. extrusion, forging and rolling so as to get isotropic matrix besides usual polycrystalline diamond tooling. Continuous reinforced technique embeds monofilament wire/fiber of boron, carbon and silicon carbide in assured path and yields anisotropic arrayed metal composite, while discontinuous reinforced technique uses fuzzy short fiber/particles of alumina and silicon carbide. The high temperature treatments are needed in fabrication of metal involved matrixes developments in order to obtain the best dispersion of constituting fiber skeletal interfaces (as on cooling yields residual strain amongst metal and reinforced fibers being vital for best composite formations [4–6]. This controlling residual stresses notably manipulates mechanical instincts of fabricated metal composites. Many metal matrix/composite have two constituents: one as metal and the other may/may not be metal or may be ceramic and/or organics. If three or more metals are used, the resultant matrix is termed as hybrid. Such metal-based matrixes/composites are complementary to heat-resistant materials like ceramic and sintered metal. Hybrid composites are innovative fiber-reinforced matrix acquired through mixing of two/ more fibers imparting improved features than other composites. Reinforcement of polymeric resin with pretty firm and low-density materials like carbon and glass fibers yields sturdy/tougher composites besides superior resistive plastic hybrid composites [15]. Certain hybrids/composites found to possess aligned and thoroughly amalgamated fiber layer besides mutually alternated depositions impart anisotropic properties owing to finally harmonize template phases. Some hybrid composites are foremost in their typical applicability in constructing lightweight structural units, orthopedic components, sturdy transporting templates used in aerospace, marine goods, sport items and trivial stuff infrastructures in building constructions [15–18]. Lightweight military aircrafts and helicopters are made with such rationally designed hybrid composites offering 20–40% reduced weights than contemporary materials. Glass fiber-reinforced carbon fiber yields hybrids/composites to be used to make rotor blades of helicopters due to innate superior fatigue resistivity as needed in making futuristic hypersonic fighter planes and aircrafts [1, 19].

4.4 Structural composites

Structural composites are obtained through geometrically designed structural elements in homogeneous pattern derived through constituents. Laminar and sandwich plates come under the category of structural composites. The strength properties of advanced structural composites offer broad mechanical properties as controlled by many parameters like volume/weight proportions of reinforced fiber/matrix components, built-up formulations, constituent mechanical features and orientations via uni- or bidirectional, besides various off-axis directional/random, arrangement. Sandwich panels are designed as lightweight structural composites owing to their comparatively elevated mechanical strengths. Such sandwich configured composites are very unique as fabricated via attachment of two thin and rigid skins to yield lightweight but bulky core slotted panel own duel outer face of relatively stiff and strong template like metal alloy, fiber-reinforced plastics, steel, and plywood adhesively bonds to thicker light-mass inner hub materials. Sandwich panels consist of inner core material made up of lightweight and low elasticity modulus like polymeric skeleton phenolics, epoxy, polyurethanes, wood and honeycombs [15, 20], while outer sheets in sandwich panels consist of tough/rigid materials so as to communicate high mechanical strength under high tensile/compressive strain loading. Sandwich panels-based structural composites offer wide utilities including in buildings' roofs, floors and walls, besides being used in fabrication of wings, fuselage and tail plane skins of aerospace and aircrafts [21–23].

4.5 Laminar composites

Some laminar composites yield through single layer fiber laminated mutual bonding or stacking own accordingly paved orientations of latent directional fluctuations achieved via consecutive depositions. The particulate platelet or laminar matrixes possess two long dimensions, e.g., wooden thin layer plywood with consecutive layers that are quite isotropic composites due to dissimilar grain/fiber orientations that are weaker in any direction than it would be if constituting fibers could all be aligned in one direction [1]. Layers of assorted fiber reinforcement yield hybrid laminate revealing anisotropic and directional structures. Based on stacking order of each layers, laminated composites owing to their in-plane and out-plane bend-stretch coupling ultimately give in-plane loading. In fact, laminar composite has two-dimensional panels or sheets with favored directions to attain highest strength [6–8]. Wood and plywood material is basically a laminated composite holding constant reinforced and preferred directional stack layer of fiber orientation instead of adhesive joints. Such laminated composite layering fetches each grain at 90° angle with its neighbors. Laminated composite attains superior mechanical strength, stability and appearance as assembled via heat, pressure and adhesive treatments. Assorted laminated composites are obtained, depending on constituents and the processing applied in their manufacturing [1–4, 6–12, 14, 15, 24]. Certain plastic-laminated glassy-type composites behold tight fit adhesiveness at their solid countertop surfaces, which were found to protect the particleboard. Cellulosic templates appear good substrates for assorted matrixes as obtained through overlay thermo-processing, e.g., laminated composite panels, medium density fiber-boards, decorative foils, high-pressure decorative composites, wood/multi-laminar veneer and resin-saturated decorative papers. Recently, numerous layer-wise dimensionally organized products like special glue/laminated timber composites are developed owing to mutual lumber bonding viable for durability, water-resistance and structural adhesiveness in resultant products, e.g., glulam a versatile stress-engineered wood beam composed of special laminations [1].

4.6 Lignocellulosic composites

Environmental causes urge to develop green composites based on renewable sources like biopolymers as economic options for glass/carbon fiber-derived composites [25]. Thus, plant-, jute- and sugar cane-based lignocellulosic composite-derived matrixes can serve this purpose. Fabrication of organic-based nanocomposites arose as a multidisciplinary area in advanced nanotechnology, particularly procured through sustainable and eco-friendly resources and methods. Some green composites are also obtained via reframing natural/bio-polymeric framework via amalgamating other natural/synthetic material substrates that offer morphological/interfacial design characteristics in resultant products over conventional counterparts [26]. Assorted biopolymeric skeletons, viz. starch, alginate, dextran, carrageenan, chitosan and cellulose, are formulated/envisaged due to their innate functional features like nontoxicity, biodegradability and biocompatibility [2, 3]. Natural lignocellulosic fibers have semicrystalline cellulose microfibril orientation offering multifunctional nanotechnological fabrications so as to cater to advanced applications in S&T as mentioned below:

4.7 Cellulose-derived nanometal matrixes

Cellulose is an extensively copious natural polymer and component of the "plant's skeleton" that exists on Mother Earth [1]. In fact, cellulose shares the same chemistry and molecular structures but imperative morphology and mechanical variations based on its recovery from various sources like vegetable, plants and bacteria. Nanocomposite matrixes are derived through cellulose skeleton via amalgamation with different phases of organic and inorganic materials [1, 13]. Nanofibrillated and bacteria-derived cellulose is made up of nanodimensional fibers, which impart novel and improved native qualities in resultant nanocomposites than vegetable-derived cellulose [13, 27]. Cellulose has soft matrix to accommodate inorganic/organic materials via blending diverse fillers to yield composites owing to their inherent functionality of constituents besides transporting unique functions due to biointerfacial alteration [7, 27]. Various material like nano-metals including gold silver and copper and inorganic gets easily doped/filled in the cellulose skeleton and yield composites which own altered fibers interaction at surfaces over bulk analogues besides coalition of fillers is beneficial for improvement of opto-electronic/electrical and mechanical functions [1, 2]. Cellulose has an exclusive arrangement and discrete affinity to form intra-/intermolecular bonding, which compacts its supramolecular ester/acetate and ether as dynamic derivatives utilized in coating, pharmaceutics, food and cosmetic industries [12, 27]. Hybrids of cellulose procured through nanometal/metal-oxides yield assorted nanocomposites like *Nowa-74* used as calorie-free dessert in food and wound dressing scaffolds in biomedical and optoelectronics. Water filtering nanomembranes are reinforced through 3D cellulose matrix owing to its brilliant characters like high purity, high polymerization degree, elevated crystallinity, high elasticity and mechanical stability and huge surface area. Certain physicochemical

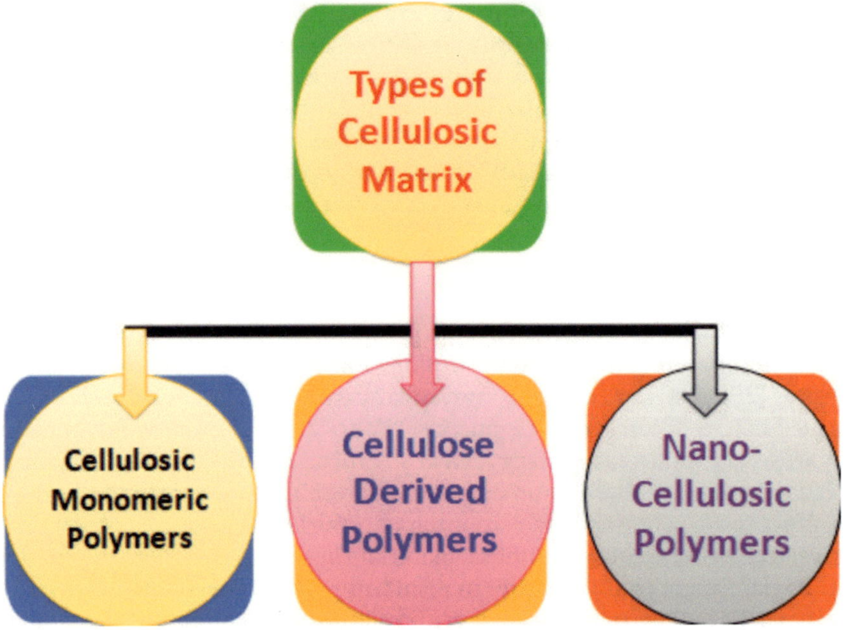

Figure 3.
Types of cellulosic matrix [1, 2, 27].

adaptations are facile to reinforce in nanofibrillated cellulose matrix so as to yield multipurpose paint additives and lacquer/latex. Reconfigured cellulose composites are obtained through inorganic nanoparticles/metal aggregates and/or other polymer blending via techniques of homogenized aggregation and in-situ metal-salt reduction in suspensions. Macromolecular networkings of cellulose matrix can be developed via facile templates and ensue nanometal distributions into its skeleton without further aggregations [13, 26–28]. Carbon reconfigured nanocomposite has received much attention due to reinforcement in mechanical features besides boosted electrical conductivity, which is additive for automotive fuel stroke components entailing electric conductance. The types of various cellulosic matrixes are depicted in **Figure 3**.

Cellulose attracted considerable attention as the strongest potential feedstock for bio-based polymer productions [1, 27]. Thus, noteworthy reinforcements in different cellulosic matrixes are done to get assorted nanocomposites. Cellulosic derivatives act as prominent filler/matrix in getting resourceful biosustainable options to high-quality synthetic polymeric composites besides substituting most petroleum-derived functional counterparts. Eco-sustainable cellulosic-reinforced nanostructures offer prospective functions with a wide applicability from energy-storage devices to biomedical scaffolds [1, 2, 13]. Cellulose is the most plentiful natural polymer and component of cell walls of plants and is also found in diverse genus, viz. algae, fungi, bacteria and sea animal/tunicate. Cellulose substance contains linearly placed and alternate stereo-configured units of D-anhydro-glucopyranose homo-polymeric bonding with 1, 4-β-glycosidic aggregations that exist as micro-fibrils [1, 2]. Morphological variations in cellulose control the degree of polymerization, which varies as per its resources. Huge hydroxyl groups on the glycosidic chain of cellulose via hydrogen bonding upshots manifold cellulosic microfibrils owing to elevated mechanical strength, rigidity, stability and biocompatibility. Hydroxyl functionality of cellulose skeleton is facile to undergo assorted physicochemical reinforcement like etherification, carboxy-methylation, cyanoethylation and hydroxyl-propylation yielding assorted derivatives for viable

DOI: http://dx.doi.org/10.5772/intechopen.91305

Cellulosic composite	Cellulose functions	Cellulose-reinforced composite property
Cellulose-lyocell/acetate, butyrate composites	Matrix/ filler	Increases tensile strength, dimensional stability, matrix compatibility, biodegradability.
Cellulose-lyocell fiber/PLA	Filler	Unexpectedly high biodegradability, significantly high mechanical characteristics.
Ethylated-cellulose, hydroxypropylated-cellulose, polyacrylated-cellulose and calcium phosphated-cellulose composites	Filler	Increases thermal and mechanical performance.
Cellulose fiber/polystyrene composites	Filler	Increases flexural storage modulus and processing speed.
Cellulose fiber/high-density polyethylene composites	Filler	Improves thermal and mechanical properties.
Cellulose particles/chitosan composite film	Filler	Enhances mechanical properties and adsorption capacity of chitosan film.
Regenerated cellulose film/BiOBr composite	Matrix	Cellulosic film own cavity for BiOBr particles and expanded specific surface area via porosity feasible for efficient photocatalysis.
Cellulose/MMT clay composite films	Matrix	High-strength cellulose composite films with excellent antibacterial activities.
Cellulose film/graphene oxide composite	Matrix	Superior mechanical performances and excellent ultraviolet-shielding properties.
Cellulose acetate/hydroxyapatite mineral composites	Matrix	Reinforced matrix holds tough HaP-cellulosic interactive ductility use to remediate pollutant.
Carboxymethyl cellulose/carbon composites	Matrix	Reinforced cellulosic composites own huge potential as sensors in bioelectronics.
Cellulose paper/carbon nanotube film/composite	Matrix	Reinforced composites are flexible, tough, thermally stable and own uniform electrical conductivity suitable for advanced biotech use.
Methylcellulose/keratin hydrolysate membranes	Matrix	Protein and polysaccharide reinforcement improves mechanical and thermal properties.
Cellulose fibers/iodine composite	Matrix	Cellulosic composites enhance photo-induced conductivity.
Cellulose acetate-polyaniline-derived membrane	Matrix	Cellulosic membranes enhance conductivity mechanical biocompatibility.
Polyhydroxybutyrate/ethyl-cellulosic-derived films	Filler	Polyhydroxybutyrate/ethyl-cellulosic reinforced formulation reduces crystallinity, promotes degradation and boosts physicochemical characters viable for sustainable biocompatibility anticipated in biomedicals and coatings.

Cellulosic composite	Cellulose functions	Cellulose-reinforced composite property
Polyhydroxybutyrate/polyhydroxyalkanoate-cellulosic composites	Filler	Cellulosic reinforcement boosts physical-mechanical strength as suitable for packaging.

Table 2.
Assorted cellulosic formulations as filler for making composites [1, 2, 13].

nanocomposites. Specific chemical formulations like ester-acetate and ether-methyl/carboxy-methylation in cellulosic hydroxyl functionality are augmented characteristics aiding in rational reconfiguration of advanced nanostructures, so it is preferred over cellulose feedstock [18, 21–23].

Visco-processing of cellulose matrix is done through derivative formation and without de-polymerization, which yields valuable products like cellophane and nitrocellulose. This cellophane acts as transparent sheet for low permeation purpose, and nitrocellulose is and excellent feedstock forming basis for rayon: the first "artificial silk" since many decades. Cellulosic derivatives were used as good feedstock for making thermoplastics since the beginning of nineteenth century. Cellulosic-reinforced formulations like cellulose acetate and cellulose esters are facile to mold as extrusion and films, besides being used in making construction materials, paints, pharmaceutical scaffolds and biodegradable plastics [1]. Cellulose-derived polymeric composites are obtained through integrating nanocellulose into assorted synthetic/natural polymeric matrixes as current advanced materials owing to their extensive applicability. Since the 1980s to the present, R&D led several innovative cellulosic reinforcements, which perk up glycol-polymeric insertion so as to yield superior functional biocomposite-derived cellulose [28]. Assorted cellulosic formulations obtained through reinforcement in host matrixes are mentioned in **Table 2**.

4.8 Advances in reinforced cellulosic nanomaterials

Currently, cellulose biopolymer can be integrated into two types of polymer nanocomposites: nanocellulose-based nanopolymer composites and nanocellulose platform-based nanocomposites. Viable applications of reconfigured cellulosic nanocomposites are shown in **Figure 4**:

Advanced biotechnology utilizes assorted biopolymer cellulosic forms including natural fibers, nanocellulose and cellulose derivatives to undergo characteristic diversified and sustainable functional variations to yield alternative composites for multifunctional usages. Copious cellulose acts as foremost natural feedstock option for fossil resources in fabricating reinforced martial matrixes [1]. Polylactic acid reinforced in cellulose matrix offers biodegradable, exceptionally sturdy and non-toxic nanocomposites called cellulosic bioplastics with superior thermal, electrical and mechanical features. Cellulosic bioplastics and functionalized nanocrystals are benign, inexpensive and robust serviceable composite owing to their well-reinforced structures of constituting matrixes for deriving electrochemical and energy-storage tools. Poly-hydroxy-alkanoate integrated/filled cellulosic matrixes yield eco-friendly composites as a substitute to synthetic polymers for food packaging, plastics and biomedical templates [1, 2]. Cellulose-reinforced composites cater to today's challenges via development of sustainable and green products through economic, environmental and social perspectives, though cellulosic biopolymeric composites fulfill partial confronts, which are to be tackled in futuristic R&D. Fully compatible two-phase polymeric composites need to be developed through materials science and process engineering in cellulose chemistry [26].

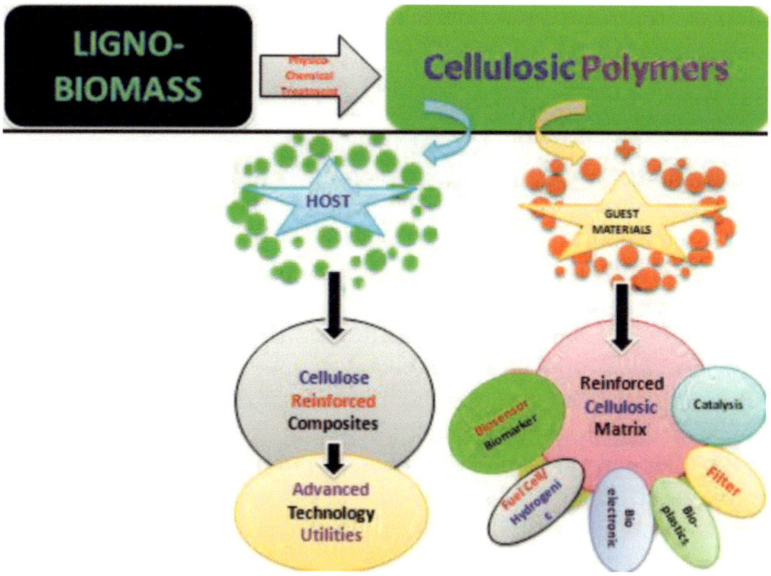

Figure 4.
Viable applications of reconfigured cellulosic nanocomposites [1, 2].

4.9 Cellulose nanocomposites

Cellulose is the copious bioproduct through plants, animals, bacteria and flora-fauna. Its extensive linear chain polymer is composed of 1,4-β-linked D-glucopyranose assembly in the hierarchy of microfibrils with excellent strength and stiffness. Nanoscale/dimensional cellulosic matrixes are reconfigured in assorted forms like nanocrystal, nanofibers or flakes. Nanocellulose matrixes appear safe, versatile, biodegradable and biocompatible without any side effects on health and environment. Reconfigured cellulose matrixes own small thermal expansion coefficient, huge aspect ratio, and superior mechanical, optical and tensile strength features. Thus, they are preferred for special utility including thermo-reversibly tenable hydrogel, paper making, coating, additives, food/drug packaging, lithe screens/films and lightweight ballistic protection beside usages in automobile windows. Assorted cellulosic composites are reconfigured, viz. dispersed phase nanofillers, dispersed phase matrix, and interfacial region hybrids owing to their potential biomedical significance, namely targeted drug/gene/cell delivery/carrying and fabricating temporary implants with PHB sutures besides making stents. Cellulose skeleton is fragile to reinforce with diverse nanometals via innate hydrogen bonding so as to yield supramolecular nanoclusters as best utilized in textiles due to native antimicrobial, antibacterial and improved catalytic parameters [26]. Nano-cellulosic reinforced polymeric matrixes owe undo recombined dynamic covalent mechanophoric linking which imparts self-healing capacity due to surface modification via scissile chemical bonding [1]. The reinforcement of material phases and skeletal faces is found to boost its surface activity in the ensuing matrixes, and due to self-supported healing, it further gifts superior sensitivity to mechanical stress transports [2]. Typical nano-cellulose composites have compacted lingo-cellulosic biomass and alter its innate characteristics like fibrils have crystalline features and high strength and mechanical rigidity. Certain nano-cellulose-based matrixes have especial features like light-weight, highly dense (1.6 g/cc) and lofty tensile strength (10 GPa at par with cast iron), e.g., nano-cellulosic matrix with proactive hydroxyl functionality provides designed and desired applications in S&T [28].

4.10 Dendrimer-reinforced polymeric matrix

Dendrimer architectures via click chemistry and self-assembly have put forward major reinforced polymeric matrixes. Advanced nanotechnology has discovered new nanoperiodic system through dendrimeric reinforcements in the form of nanodevices, nanomaterials and nanomedicines. This decade has developed many scientific and commercial utilities focusing architecturally on drive characteristics through assorted dendrimeric reinforcements. Dendrimeric reinforcements convey highly defined and well-branched perfect nanostructures obtained through repetitive branched monomer iterative protection and deprotection in contrast to hyperbranch polymeric skeletons. Rationally reconfigured dendrimeric/dendritic polymers are used for drug delivery and catalysis and making light-emitting materials [1, 2]. Today, dendrimer chemistry and catalog of dendrimers offer templates for organic-inorganic hybrid nanomaterials imparting diversified and captivated utility in modern S&T. Hyperbranch polymers are 3D globular low viscous and highly soluble template used in making dendrimer-based organic/inorganic nanohybrids. Organic-inorganic nanohybrids are reconfigured through hyperbranched dendrimeric matrix to get mono-/bimetallic, bimetallic alloys and core/shells. Dendritic hyperbranch polymers and dendrimers with 3D globular arrangement and spherical outline are found to offer high branching density and branching at each repeating unit, which imparts exclusive features like lower viscosity and high solubility/functionality than linear counterparts. Advanced dendritic matrixes blend reconfigured organic/inorganic frameworks via good skeletal processing, thus offering special developed advantageous features like superior elasticity, light-weight, impassive resistance, robust strength, chemical resistance, and thermal stability. Reinforced hyperbranched dendritic matrixes own linear segments of globular end functionality valuable for making many organic-inorganic nanocomposites [2]. Assorted nanohyperbranched composites with unique globular contour with functional end groups are obtained through organic-inorganic dendrimeric templates like polyamidoamine-reinforced nanogold particles, nanocarbon hybrid sols and heterogeneous nanostructures [2]. Intrinsic void-reinforced nanoparticles and metal clusters with enhanced stability are exploited for catalysis, adsorptions and photodetection besides being used in developing antimicrobial sensors/agents [1]. Recent advancement in nanotechnology aids to induce "disordered-to crystalline" features and perform numerous structural reinforcements a way from metal aggregate to nano-crystals in the resultant clusters/matrixes that are especial for electro-catalytic usages [1, 2]. Reinforced carbon black matrix is used for air CO_2 absorption/desorption better than Excellion™ ion-exchange membranes.

5. Controlled practices for reconfiguration

"Click" chemistry is found to control reconfigurations or modifications and facilitates stimulus detachment through characteristic porous dendrimeric links so as to yield assorted structures like fine layer-by-layer films, nanoparticles, nanosheets, nanowires and nanotubes especially used for optoelectronic nanodevices [2]. Reinforced dendritic architectures with linear, cross-link and chain-branching can afford many features like open-space functionalization and organize topology, copolymeric hybridization and terminal grafting. Versatile dendritic reinforcement motivates innovative R&D beyond predicted applications in many industries. Hyperbranch supramolecular polymer-based reinforced material matrixes offer sophisticated chemical and biological utilities. Nonmaterial reconfigurations and nanostructure reinforcements are pivotal themes in today's scientific and technology advancement. Rationally developed nanomaterials/structures have vividly amended characteristics viable for assorted utilities, viz. electric tools, optoelectronic devices, biosensor/biomarker, photodetector, solar cell, quantum dots and plasmonic modules. At nanodimensions, the interfacial phase interaction gets better without changing matter itself; instead, it has shown innovative properties like electrical conductivity, insulation, boosted reactivity, elasticity and superior/robust strength, which is missing in micro-/macroscale counterparts [1–4, 6, 7]. These nanocomposites have multiphase combination of two/more components and one or more fillers of particles, sheets and/or fibers in their reinforced matrixes. Certain nanocomposites are reinforced through numerous progressively designed alterations in material matrixes and novel comprehensive techniques thrust interdisciplinary R&D in chemistry, physics, biology and biotechnology besides paving the way for business breakthroughs in current S&T developments [1]. Reinforced building blocks with nanodimensions fetch practical and theoretical interests via rational reconfigured and designed innovative nanocomposites owing to their extraordinary physicochemical characters and preset functionality [2]. R&D findings through inventive analytical trends and developments in nanocomposite chemistry with compiled data and analysis may aid in the exploration of stimulated reinforcements in nanocomposite matrixes [26, 28].

Nanocomposite/nanostructures reconfigured through biomatrixes faced intricate tasks like choice, combinatorial paths, modified synthesis and shaping of proper materials. Thus, plant tissues like cellulosic matter and animal components like bones/cartilage are reinforced at nanodimensional hierarchy. Morphological likeness of lignocellulose, hemicellulose and lignin matrixes enables its vast reconfigurations in developments of nanostructures/composites, which acted as viable alternatives in modern drug delivery and tissue engineering scaffolds (for regeneration of bones and cartilages) besides being used in cell-attached proliferated inductions. Biomimetic composites are being reconfigured through reinforcement of natural polymers like chitin and cellulose with inorganic materials like calcium phosphate for use in biomedical applications. Advanced nanotechnology paves the way for prospective capability across a broad spectrum of applications. The advancements in the field of nanotechnology have generated many reinforced material blends, alloys, matrixes and composites offering contemporary functional applications over other known counterparts [1].

Nowadays, specially sophisticated designing and targeted tailor/engineer morphologies are facile to induce at any (0D, 1D, 2D, and 3D) nano-dimensional scales with structure-property dependent remarkable parameters and controllable interfacial area, e.g., many onset non-bulky size-dependent quantum dots and polymeric nano-composites for industrial and environmental applications [1, 2]. Rationally reinforced polymer-based nanocomposites are best options to usual polymeric fillers and blends that appeared as a staple part of modern plastics. Nanotechnological frontiers in the twenty-first century seek better optimized composite combinations with innate synergistic utility; thus, material and functional devices need to be reinforced or reconfigured. Thus, nanotechnological reinforcements are solely dependent on mechanical augmentation of matrixes as substitute for existing counterparts. Benign or biodegradable nanocomposites reinforced through green polymers like cellulosic and chitosan matrixes are potential options to petroleum-derived non–eco-friendly thermoplastic polyolefin nanoclusters [7, 13, 27]. Reconfigured polymeric nanocomposites with exclusive mechanical, physicochemical, thermal, electrical and barrier properties and fire-retarding properties emerge as innately prospective in making exterior/interior and underbonnets, coating and components of automotive [1–4, 6, 7]. A Global Strategic Business

Report stated that the world's nanocomposite automotive market may exceed over one billion pounds in this decade with an ever robust demands in futures. Green nanocomposites obtained via reinforcements of clay, nanocarbon and other nanofibers are beneficial than usual counterparts due to boosted mechanical, electrical and thermal barriers, besides high tensile force, more deflection temperature and flame retardations [1, 29]. Certain polymer-derived nano-material matrixes are found to possess unchanged innate features, viz. native power, viscosity and parallel optic potential by virtue of legitimated morphological reinforcements comprising thousands of hoard layers at nano-scale ensuing exfoliation and dispersion all over surfaces [2]. Reconfiguration further upshots degree of exfoliation in resultant nanostructures ultimately offering greater surface area with improved performance [2, 19, 20].

6. Challenges in designing/reinforcement of composite/matrix

Certain major challenges of composite materials are mentioned as follows:

- Need systematic and progressive interactive engineering approaches for applying various methodological technologies so as to yield optimal material characteristics with aiming targeted composite/hybrid matrixes.
- Require critical investigative analysis of physical, chemical and mechanical properties of each constituent and resultant composites/hybrids.
- Necessary to formulate some properties in resultant composite/matrix as optimized and precise with respect to its operational conditions.
- Focus to comprehend consequential manufacturing and compositional parameters of composite/matrix.
- Aim to develop meticulous and progressive combinatorial computational techniques with optimized study for expected efficiency/capacity at varied functioning conditions.
- Target to epitomize robust inhomogeneous properties in procured significant composite.

7. Prospective progress of rationally reinforced matrixes

Several challenges needs to be tackled during matrix reconfigurations and nano-composite fabrication/progressing, and this allows progressive rational designing as per desired applications. The chemical interactions that exist in resultant matrix reinforcement during its fabrication and later in its usage are to be checked via micromechanics to provide a rational basis for its designing. Intrinsically strengthened composites are obtained through various reinforced materials like graphene, nanocarbon and nanometals, which are found superior than counterparts. Reconfigured multiscale hybrids own better load reassignments at reinforced interfaces achieved via tailored interfacial shear strengthens imparting stiffness, heat protection, impactive compression and better threshold load bare. Various automotive utility needs reconfigured biopolymer-derived nanocomposites at lower cost and end-use components.

8. Innovative breaches in applicability of reinforcement materials

Some innovative breaches in applicability of rationally reinforced materials are observed that are mention as follows:

- Nanometal reinforcement in cellulose matrix is attenuated through synergy of constituting materials so as to derive innate advantageous properties of components by diverse preparatory techniques.
- Reconfigured metal nanomatrixes of chemically integrated cellulosic templates with certain novel functionalities in the ensuing composites are best exploited for filler purpose. Such nanometallic reconfigured cellulose-derived matrixes with multifaceted features have not been explored extensively and thus need to be attended.
- The basic physicochemical interactions found in such components that viably optimize innate properties are less studied.
- Nanocomposite formations significantly alter and improve the innate properties of components via multiphase reinforcements.
- Advance nanotechnology aids in reconfiguration of various materials into sole composite imparting advantageous and optimized features as per the industrial needs.
- Assorted constituting textures and dimensional comprising phases of host matrixes via their innate bonding control mechanical stability and properties of resultant nanocomposites, e.g., nanometal/metal oxide dispersions in ceramics or vitreous.

Acknowledgements

The author is thankful to Head, PGTD of Chemistry, R.T.M. Nagpur University, Nagpur, for laboratory facilities and to the Vice Chancellor, Nagpur University, Nagpur, for the sanction of research project work under University Research Project Scheme, awarded vide No. Dev/RTMNURP/AH/1672 (9) dated September 24, 2016.

Author details

Rajendra Sukhadeorao Dongre
Department of Chemistry, R.T.M., Nagpur University, Nagpur, India

*Address all correspondence to: rsdongre@hotmail.com

References

[1] Dongre RS. Rationally fabricated nanomaterials for desalination and water purification. In: Book Novel Nanomaterials. Vol. 1. Croatia: In-Tech Open Publisher; 2018. pp. 348-366. DOI: 10.5772/intechopen.70149

[2] Dongre RS. Biological activities and application of marine polysaccharides. In: Marine Polysaccharides in Medicine. Vol. 1. Croatia: In-Tech Open Publisher; 2017. pp. 181-206. DOI: 10.5772/65786

[3] Yi X-S, Shanyi D, Zhang L, editors. Polymer matrix materials. In: Composite Materials Engineering, Volume 1: Fundamentals of Composite Materials. Singapore: Springer; 2017. pp. 147-168. ISBN: 978-981-10-5696-3

[4] Luo YF. New developments in hi-tech synthetic fibers. Hi-tech Fiber & Application. 2000;**25**(4):1

[5] Kotek R. Recent advances in polymer fibers. Polymer Reviews. 2008;**48**(2):221

[6] Shaghaleh H, Xu X, Wang S. Current progress in production of biopolymeric materials based on cellulose, cellulose nanofibers, and cellulose derivatives: Review. RSC Advances. 2018;**8**:825-842. DOI: 10.1039/C7RA11157F

[7] Huang X, Zheng S, Kim I. Hyperbranched polymers & dendrimers as templates for organic/inorganic hybrid nanomaterials. Journal of Nanoscience and Nanotechnology. 2014;**14**(2):1631-1646. DOI: 10.1166/jnn.2014.8750

[8] Ebrahimi F. Nanocomposites: New Trends & Developments. Croatia: In-Tech; 2012. DOI: 10.5772/3389

[9] Komarnenei S. Nanocomposites. Journal of Materials Chemistry. 1992;**2**:1219-1230

[10] Prud'homme R, Ozbas B, Aksay I, Register R, Adamson D. Functional Graphene? Rubber Nanocomposites. U.S. Patent 7,745,528, 2010 (Filed 2006). Assignee: The Trustees

[11] Netravali AN. Green composites: Current trends and developments. In: Proceeding of MACRO-04. Turuvanthapuram, India; 2004

[12] Klempner D, Sperling H, Utracki A. Interpenetrating Polymer Network. Washington DC, USA: American Chemical Society; 1994

[13] Barham PH, Leller A. The relationship between microstructure and mode of fracture in polyhydroxybutyrate. Journal of Polymer Science: Polymer Physics. 1986;**24**(1):69-77

[14] Leo Baekeland. "Plastics. UK history site. 28 June 2000. New Chemical Substance" (PDF). The New York Times. Princeton University; 06 February 1909

[15] Netravali AN, Chabba S. Composites get greener. Materials Today. 2003;**6**:22-29

[16] Chabba S, Netravali AN. 'Green' composites using modified soy protein concentrate resin and flax fabrics and yarns. Japan Society of Mechanical Engineers. 2004;**47**(4):556-560

[17] Chamis CC. In: Piuddemann EP, editor. Interfaces in Polymer Matrix Composites. New York: Academic Press; 1974

[18] Piggoti MR. Relations between and fiber-polymer interface properties. In: Proceedings of ASC 6th Technical Conference. 1991. pp. 725-731

[19] Davis ME. Ordered porous materials for emerging applications. Nature. 2002;**417**:813-821. DOI: 10.1038/nature00785

[20] Kiba S, Suzuki N, Okawauchi Y, Yamauchi Y. Prototype of low thermal expansion materials: Fabrication of mesoporous silica/polymer composites with densely filled polymer inside mesopore space. Chemistry, an Asian Journal. 2010;**5**:2100-2105

[21] Bănică F-G. Chemical Sensors and Biosensors: Fundamentals and Applications. Chichester, UK: John Wiley & Sons; 2012. p. 576

[22] Anderson M, Tilman AM. Acetylation of jute: Effects on strength, rot resistance, and hydrophobicity. Journal of Applied Polymer Science. 1989;**37**(12):3437-3447

[23] Sanadi AR, Young RA, Clemsons C, Rowell RM. Recycled newspaper fibers as reinforcing fillers in thermoplastics: Part I - Analysis of tensile and impact properties in polypropylene. Journal of Reinforced Plastics and Composites. 1994;**13**(1):54-67

[24] Goriparthi BK, Suman KNS, Rao NM. Effect of fiber surface treatments on mechanical and abrasive wear performance of polylactide/jute composites. Composites Part A: Applied Science and Manufacturing. 2012;**43**(10):1800-1808

[25] Okubo K, Fujii T, Yamashita N. Improvement of interfacial adhesion in bamboo polymer composite enhanced with micro-fibrillated cellulose. JSME International Journal. 2005;**48**(4):199-204

[26] Graupner N, Herrmann AS, Mussig J. Natural and man-made cellulose fibre-reinforced poly(lactic acid) (PlA) composites: An overview about mechanical characteristics & application areas. Composites Part A: Applied Science and Manufacturing. 2009;**40**(6-7):810-821

[27] Nakagaito AN, Yano H. Novel high-strength biocomposites based on microfibrillated cellulose having nano-order-unit web-like network structure. Applied Physics A: Materials Science & Processing. 2005;**80**(1):155-159

[28] Silva MJ, Sanches AO, Medeiros ES, Mattoso LHC, McMahan CM, Malmonge JA. Nanocomposites of natural rubber and polyaniline-modified cellulose nanofibrils. Journal of Thermal Analysis and Calorimetry. 2014;**117**:387-392

[29] Nakagaito AN, Iwamoto S, Yano H. Bacterial cellulose: The ultimate nano-scalar cellulose morphology for the production of high-strength composites. Applied Physics A: Materials Science & Processing. 2005;**80**(1):93-97

Chapter 7

Composite Nanofibers: Recent Progress in Adsorptive Removal and Photocatalytic Degradation of Dyes

Duy-Nam Phan and Ick-Soo Kim

Abstract

This chapter intends to review the state of the art of a new nanomaterial generation based on electrospun composite nanofibers for dye removal from wastewater. Natural polymer-based nanofibers, nanofibers with unique morphology, and carbon nanofibers were comprehensively reviewed as capable carriers for a broad spectrum of functional materials such as metal oxides, zeolite, graphene and graphene oxide (GO), and metal-organic frameworks (MOFs) in the application of dye removal. The various nanostructures, adsorption capacity, advantages, and drawbacks were discussed along with mechanistic actions in the adsorption process and photocatalytic performance that emphasize current research development, opportunities, and challenges. The chapter covers multiple intriguing topics with in-depth discussion and is a valuable reference for researchers who are working on nanomaterials and the treatment of colored waters.

Keywords: composite nanofibers, electrospinning, dyes, adsorption, photocatalytic degradation

1. Introduction

The activities of textile, printing, leather, paint, and paper industries are discharging millions of gallons of wastewater every day, contaminating water bodies and terrestrial lands. The impacts on the environment are irretrievable and gravely dangerous. The organic dyes in effluents and discharges used for fabrics and colored materials are persistent in water and pose long-term effects on human health, fish, and aquatic organisms. Many synthetic dyes are considered toxic, carcinogenic, and mutagenic; even a small amount infiltrates the human body. Direct contact with dyestuffs can lead to skin allergy and neurological, reproductive, and endocrine diseases [1].

Nanofibers with high surface area to volume ratio, excellent flexibility, porous structure, reusability, nontoxicity, environmental stability, and low cost are suitable supporting materials for loading functional materials or being modified with different chemical groups in water treatment application. With enhanced surface area and pore volume, the nanofibers as a filtering media deliver high contact between adsorbent and aqueous media, resulting in improved adsorption capacity with the convenience of recovery and recycling. By engineering various functional groups (carboxylate, amino, acid, and hydroxyl groups) or the integration of adsorbents, including metal oxides, graphene, graphene oxide (GO), and metal-organic frameworks (MOFs) in the nanofibers, the separation capacity can be greatly improved [2–4].

Among various systems, which have been developed for the removal of dyes in wastewater, namely, adsorption, ion exchange, membrane filtration, and coagulation, adsorption is the most effective and versatile strategy to remove dyes at high concentrations with high removal percentage. The adsorption process involves several stages: (i) dissolving dyes into the solution, (ii) the external diffusion of dyes to the surroundings of the adsorbents, (iii) internal or intra-particle diffusion which fills nanoparticle pores with dye molecules, and (iv) adsorption or desorption on the interior sites. If the amounts of dye uptake are correlated with the square root of time in a linear relation, the adsorption process is significantly influenced by intra-particle diffusion because step (iv) usually happens rapidly [5].

Most dyes are water-soluble and can be classified as cationic, anionic, and nonionic; the names are derived from the charging states when being dissolved into an aqueous medium. Depending on the chemical structures of dyes, the approaches and adsorption conditions can vary accordingly, which include material selection, adsorption or photocatalytic degradation, pH, time, and temperature. A spectrum of organic and inorganic materials such as transitional metal oxide, graphene and GO, carbon nanotubes, zeolites, and MOFs have been used for treating colored waters. These materials are suited for separating dyes from wastewater owing to abundance, low cost, ease of being employed, adsorptive selectivity, and biocompatibility [6–8].

The photocatalysis process has emerged as a newly developed technique for wastewater remediation. Photocatalysts with a particular bandgap can be activated by different light sources to generate electron-hole pairs, which either recombine or migrate to the surface and initiate photocatalytic reactions. After that, the holes oxidize H_2O to produce hydroxyl radical OH•, whereas electrons react with absorbed O_2 to produce oxygen radicals O_2• and other intermediate forms [9]. The hydroxyl

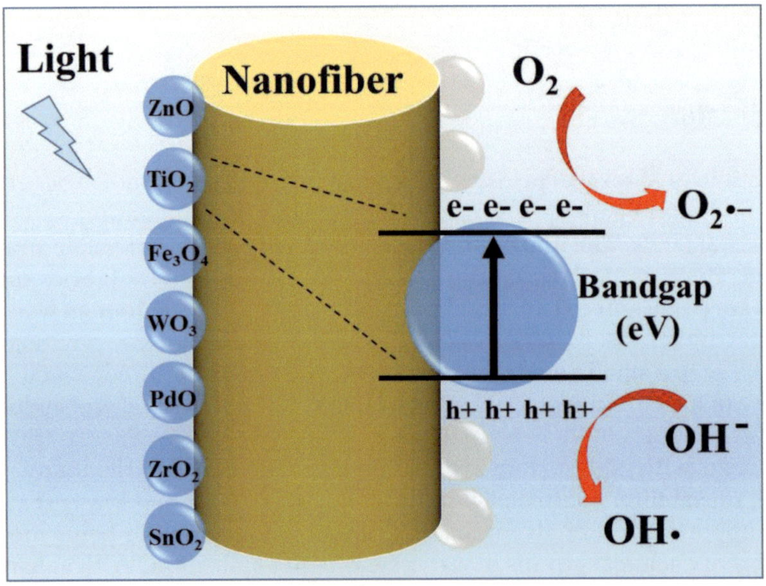

Figure 1.
Scheme for photocatalysis of metal oxide nanoparticle-decorated nanofibers under UV or visible light sources.

radicals and oxygen radicals then attack dye molecules to convert pollutants or contaminants into nontoxic forms or completely decompose them to CO_2 and H_2O (**Figure 1**). The criteria for useful photocatalysts are the capability to absorb the solar spectrum at the visible range, excellent performance, and long-term stability.

2. Electrospinning technique, natural polymer-based nanofibers, and carbon nanofibers (CNFs)

2.1 Electrospinning technique as the fabrication method

Electrospinning is one of the several well-developed techniques to fabricate fibers at micro- or nanoscale (**Figure 2**). The electrospinning with versatility allows excellent controls over the fiber diameters, nanostructures, and morphology to enhance catalytic, mechanical, electrical, biomedical, optical, and adsorptive properties. With a wide selection of polymers and the facilities for additive incorporation, the electrospinning process can manufacture nanofibers into different fascinating structures for varied applications [14]. With recent advancements in the electrospinning technique, fascinating nanostructures could be obtained with inspiration from objects in nature and can be applied in improving pollutant removal. The tree-like structure is composed of trunk fibers and branch fibers. The trunk fibers with the support role can improve the mechanical property, and the thin branches play the role of connection, decrease the pore size of the membranes, and increase the surface area [15]. The spider web-like structure was

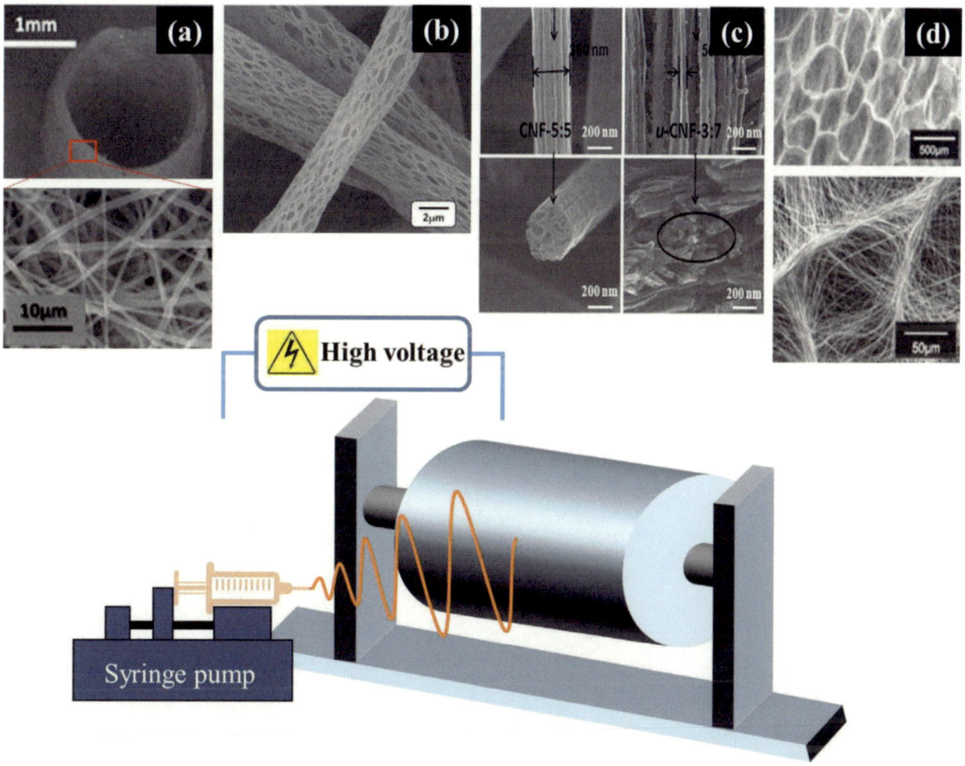

Figure 2.
Electrospinning technique to fabricate nanofibers with different morphology (a) poly(ε-caprolactone)-poly (l-lactic acid) nanofiber tubes [10], (b) cellulose acetate nanofibers with morphology control [11], (c) porous carbon nanofibers [12], and (d) cellulose acetate nanofibers with honeycomb-like surface structure [13].

fabricated by growing zeolitic imidazolate framework-8 (ZIF-8) nanocrystals on the nanofibers. The nanofibers showed high removal efficiency for incense smoke, formaldehyde, and PM particles, which was attributed to the improved surface area and electrostatic interaction between ZIF-8 and particles [16]. Hierarchical bioinspired composite nanofibers comprised of PVA, PAA, GO-COOH, and polydopamine demonstrated the eco-friendly and controllable fabricating process with efficient adsorption capacity for dye removal [17]. The excellent adsorption was due to the strong electrostatic field of carbonyl group modified GO and the unique structure of polydopamine. The membrane exhibited excellent reusability with the potentially large-scale application.

2.2 Electrospinning bio-based polymers for water treatment

Due to concerns about sustainability and environment, bio-based polymers such as cellulose, chitosan, zein, collagen, silk, hyaluronic, alginate, and DNA have been used significantly to fabricate nanofibrous composite membranes [18]. The applications of these polymers in water filtration at the commercialization scale have seen the increase over the past few years due to the beneficial properties of biocompatibility, biodegradability, safety, and nontoxicity. One of the unique features of bio-based polymer is the possession of various functional groups, which can be utilized for pollutant collection (**Figure 3I**). The dye adsorption mechanisms onto polymers can be of chemisorption or physisorption. The former is often related to strong bonding (such as covalent or ionic bonding) or chemical reactions and is irreversible. The latter is a reversible process, thus more preferable. The physisorption is governed by van der Waals forces, hydrogen bonding, hydrophobic interaction, and electrostatic attraction.

Gopakumar and coworkers modified cellulose nanofibers by esterification with Meldrum's acid, which endowed the nanofibers with the affinity toward positively charged dyes [21]; the mechanisms of adsorption were suggested as electrostatic forces between carboxylate groups and the dye molecules. Chitosan/polyamide nanofibers were reported to have excellent adsorption capacity toward anionic dyes, 456.9 mg/g for Reactive Black 5 (RB5) and 502.4 mg/g for Ponceau 4R (P4R), primarily due to the affinity of amino and hydroxyl groups in the chemical structure [22]. With the increase of the ratios of chitosan/polyamide, the adsorption capacities improved, which was assigned to the fact that more reactive sites are present in chitosan than polyamide. Li et al. reported an efficient and facile route to cover electrospun silk nanofibers with MOFs for high removal efficiency toward

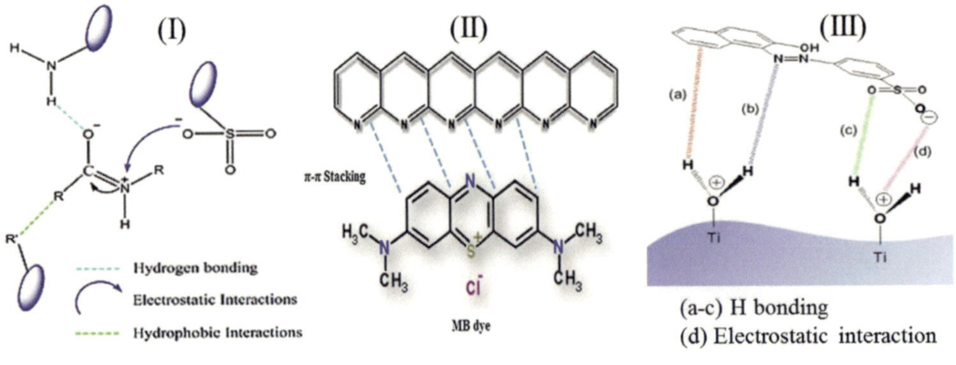

Figure 3.
Mechanism of dye affinity: (I) RB5 on zein nanofibers based on hydrogen bonding, electrostatic interactions, and hydrophobic interactions [19], (II) MB on CNFs governed by π-π stacking interactions [20], and (III) orange II on titania aerogel via H bonding and electrostatic forces [5].

rhodamine B (RB) and malachite green (MG) [23]. The authors successfully loaded the composite nanofibers with high contents of MOFs, and more importantly, the 3D structure of MOFs was well retained within the silk nanofibrous membrane.

2.3 Carbon nanofibers as supporting materials

Carbon nanofibers with unique and tunable morphology have been used in catalytic, environmental, and energy applications [24]. Polyacrylonitrile (PAN) nanofibers have been used extensively as an efficient precursor for CNF fabrication. CNFs have been employed as an efficient carrier for the loading of catalytic and bioactive materials. A broad spectrum of metal oxide nanoparticles has been immobilized in/onto CNFs for dye removal applications. The main approaches to decorate CNFs with active materials include electrodeposition, chemical synthesis, and dry synthesis. Interestingly, the CNFs also present adsorption capacity (**Figure 3II**), due to the binding between aromatic rings of CNFs and adsorbate based on π-π stacking interaction [20]. The high conductive and chemically inert properties of CNFs enhance its application in dye degradation via photocatalysis and promote reusability. The dye molecules have high chances to be attracted to CNFs, having a π-conjugative structure, before being decomposed by photocatalysts loaded on the surface of the nanofibers [25].

3. Adsorptive removal by inorganic absorbents incorporated into nanofibers

3.1 Metal oxide

Transitional metal oxide nanoparticles, including copper oxide, zinc oxide, iron oxide, titanium dioxide, and mixed metal oxide nanocomposites, have been investigated in the dye uptake or dye removal efficiency. Metal oxides have remarkable physical and chemical characteristics, which have been proven useful for water purification. The electrostatic attraction, hydrophobic interactions, and hydrogen linkages between the surface of metal oxides and dye molecules were supposed to dominate the adsorption, controlling the kinetics and isotherm of adsorbent-adsorbate interactions [26].

Li et al. reported that the maximum adsorptions for Fe, Co, and Ni oxides were found to be at neutral pH and the rise of temperature has a positive impact on the capacity of dye removal. The BET surface areas of these composite nanoparticles were reported to be between 97.26 and 273.5 m^2/g [6]. The plausible explanation for the best adsorption capacity at the neutral region is the corrosive destruction of metal oxide nanostructure at high or low pH. At acidic pH, the leaching of metal happens because of the reaction between metal oxides and H^+. At alkaline pH, the hydroxyl groups attach to the active sites on the metal oxides, which are positively charged, reducing the available number of sites and thereby the attraction between adsorbent surface and dye molecules. Malwal and coworkers reported that the pH_{PZC} values of CuO and ZnO nanoparticles were around 9.4 and 9.5, which means at pH lower than pH_{PZC}, the surface of CuO and ZnO becomes positively charged. The electrostatic attraction is the primary mechanism of anionic dye affinity [27]. Similarly, the pH_{PZC} of MgO is 12.4, and the anionic dye adsorption is driven by electrostatic forces. The pH_{PZC} of TiO_2 is around neutral values (6–6.8), which is not so much different from those of iron oxides; the PZC values measured for FeO, Fe_3O_4, and Fe_2O_3 were around 6.1–6.5 [28, 29]. The pH_{PZC} of TiO_2 and iron oxides is in the neutral range, making them efficient adsorbents for both cationic and anionic dyes in a wide range of pH mediums.

The mobilization of metal oxides into carbon nanofibers by several methods has been described in the literature to improve the adsorption performance. The traditional one is to disperse precursors of metal oxides in the PAN polymer solution before electrospinning and carbonization. Nevertheless, the metal oxide nanoparticles are usually located inside the CNFs, which resulted in low adsorption efficiency. Besides, the agglomeration is also of concern because it is detrimental to dye removal efficiency and mechanical properties of the CNFs. The ultrasonic decoration of CNFs with TiO_2 was a straightforward technique to achieve uniform distribution of nanoparticles and yield higher efficiency of dye uptake [30]. TiO_2@carbon composite nanofibers can be prepared by electrospinning technology, followed by a hydrothermal method to acquire the nanoarray structure [31]. The high adsorption performance was explained as the decoration of TiO_2 nanoarray induced the specific surface area enlargement, the tunable wettability from hydrophobicity to the hydrophilicity of the carbon nanofibers, and considerable negative Zeta potential value. Furthermore, the addition of $TiCl_4$ in the electrospinning solution increased the macroscopic flexibility and the adsorption performance of CNF from 9.92 to 24.77% for methylene blue (MB), respectively.

3.2 Zeolite

Zeolite, an aluminosilicate framework obtained from nature, can be chosen as a suitable filler material in the polymeric nanofibrous matrix due to its porous structure and exchangeable cation feature [32]. Its 3D structure with negatively charged lattice, high specific surface area, and competitive price makes zeolite an appealing choice for dye adsorption. The adsorptive sites in zeolites can be controlled by adjusting the ratio between silicon and aluminum. With its strong adsorption capacity for waste products and toxins, zeolite has been reported to show affinity toward methyl orange (MO) [33, 34], MB, and MG [35], with high adsorption capacity and reusability feature. The adsorption mechanisms are complicated, including porous structure, charged surfaces, heterogeneity, and other imperfections. Lee et al. reported that PMMA/zeolite nanofibers exhibit high removal efficiency up to 93% for MO at 30 mg L^{-1}. The isotherm adsorption results were fitted well with the Langmuir model, which indicated that the dye molecules were adsorbed onto the homogeneous surface and monolayer adsorption existed during the process [34].

3.3 Graphene and graphene oxide

Recently, graphene and GO have been studied extensively in the field of catalysis and adsorption as a result of their massive surface area, delocalized π network, and inertness to be used in a wide pH range. Graphene has features of chemical stability, low toxicity, and hydrophobicity. The oxidation of graphene provides an excellent hydrophilic surface; at the same time, it compromises the π electron structure, resulting in poorer attraction to aromatic hazards [36]. The reduction of GO, which forms rGO, is a process to recover the adsorption capacity for GO by giving it back the π-delocalized electron structure and hydrophobic property. Graphene-based materials tend to aggregate due to strong van der Waals and π-π interactions; thus, incorporating them into polymeric nanofibers is a way to overcome the aggregation [7]. Composite GO/PVDF nanofibrous membrane was prepared by ultrasonic treatment for the use of organic dye removal. The facial treatment technique, with the support of ultrasonication, was implemented. The adsorption capacity is mainly dependent on GO contents of the composite membranes, and the pseudo-second-order model showed a better fit [37]. The mechanism of adsorption was suggested for π-π stacking interaction between delocalized π electrons in graphene and aromatic rings of dyes [38].

3.4 Metal-organic frameworks

MOF is an excellent porous media with a multitude of applications in biomedical engineering, photocatalysis, CO_2 separation, and dye removal. With the properties of chemical and physical stability, effective surface area, excellent adsorption capacity, and nontoxicity, it has been widely used as an essential material for environmental remediation [39]. However, its poor processability hinders the fabrication into filtration devices. Many researchers have successfully applied MOF-based composite nanofibers for contaminant removal from wastewater. Li et al. reported co-electrospun anionic MOF nanofibrous membranes, which displayed synergistic action of PAN and bio-MOF-1 in the adsorption process for MB [8]. The resulting filter could sustain a constantly high adsorption capacity because of the stable nanofibrous structure and no leaching effects. Desorption was conducted in a saturated Na^+ solution based on the ion exchange equilibrium. The ion exchange process happened to settle the dynamic equilibrium between ions of different species. The high adsorption performance of MOF embedded in the polymeric nanofibers could be explained as the diffusion of dye molecules to the surface and internal channels of MOF, which is governed by a multilayered adsorption process associated with the transportation of Gaussian energy into a heterogeneous structure [39].

3.5 Recent novel adsorbents for dye uptake

The surface functions of electrospun composite nanofibers are crucial for dye removing applications, which depend partly on the chemical groups of the used polymers and can be modified by chemical grafting or loaded absorbents. Novel p(NIPAM-co-MAA)/β-CD nanofibers were fabricated by electrospinning and thermal crosslinking for the application of crystal violet (CV) removal. The porous structure obtained from high-temperature treatment caused a hydrophobic surface, which facilitated the dye removal. The high adsorption capacity was attributed to electrostatic attraction, host-guest interaction of β-cyclodextrin, and hydrophobic forces [40]. Zhang and coauthors synthesized acid-activated sepiolite fibers grafted with amino groups for the adsorption of Congo red (CR) [41]. The Weber and Morris model fitting suggested that the adsorption happened through two stages, which included the initial period involving the external mass transfer and the final stage governed by intra-particle diffusion.

Recently, clay minerals have been intensively studied for the fabrication of clay-polymer composite nanofibers owing to the benefits of low cost, nontoxicity, and good adsorption [42]. Montmorillonite/chitosan/PVA nanofibers were utilized for Basic Blue (BB41) separation. The complex formation between amine groups and cationic dyes governed the adsorption and gave an explanation to the maximum adsorption capacity of the composite material at a pH of 7. At acidic pH, the active sites were occupied by hydrogen ions. Natural calcium alginate with biocompatibility and nontoxicity shows promises in colored water treatment due to possessing carboxyl groups, which can attract cationic dye molecules. Gelatin with amino groups also presents high adsorption performance against dyestuffs. The combination of two materials in the form of composite nanofibers showed good adsorption capacity with improved reusability and regeneration compared to using only calcium alginate nanofibers [43].

No.	Adsorbent	Dye	Adsorption capacity, mg g^{-1}	Reference
1.	PVDF/GO nanofibers	MB	621.1	[37]
2.	MOF/PAN nanofibers	MB	20.68	[8]
3.	PVA/PAA/GO-COOH@PDA	MB	26.45	[17]
4.	Gelatin/alginate composite nanofibers	MB	1937	[43]
5.	Zeolitic imidazolate framework-8 functional polyacrylonitrile nanofibers	MB MG	36.92 1531.94	[35]
6.	ZIF-8@CS/PVA-ENF	MG	1000	[39]
7.	MOFs grew on silk nanofibers	RB MG	19 840.2	[23]
8.	CuO-ZnO composite nanofibers	CR	126.4	[27]
9.	NH_2 grafted acid-activated sepiolite fibers	CR	539.71	[41]
10.	Meldrum's acid cellulose nanofibers-based PVDF nanofibers	CV	3.984	[21]
11.	β-Cyclodextrin modified p(NIPAM-co-MAA) nanofibers	CV	1253.78	[40]
12.	Chitosan/polyvinyl alcohol/zeolite electrospun nanofibers	MO	153	[33]
13.	PMMA/zeolite nanofibers	MO	95.33	[34]
14.	Chitosan/polyamide nanofibers	RB5 P4R	456.9 502.4	[22]
15.	APAN/Fe_3O_4–MPA composites nanofiber	Indigo carmine	154.5	[45]

Table 1.
Comparison of different composite nanofibers for the adsorptive removal of dyes.

Owing to the mesoporous structure and the possibility of functionalization, meso-silica has drawn significant interest in the field of dye adsorption. The surface of meso-silica modified with carboxylic acid groups showed affinity toward cationic dyes but presented almost no adsorption for anionic and neutral dyes. The inorganic modification of meso-silica with CuO enhanced the adsorption effects on the cationic dye, which was related to electrostatic forces between CuO and dye molecules [44]. Adsorption capacities of different composite nanofibers for various dyes are listed in **Table 1**.

4. Photocatalytic degradation of dyes using composite nanofibers

4.1 ZnO-loaded nanofibers

The photodegradation is a light-induced process following the contact of contaminants to the photocatalysts, and its efficiency is substantially governed by the adsorption capacity of photocatalysts. Therefore, the adsorption of pollutants into metal oxides is the prerequisite for efficient photodecomposition, which hints that it is necessary to increase the surface area of adsorbents to give more binding sites and restrict the aggregation. Reducing the sizes of metal oxide to nanoscale and loading them onto the surfaces of nanofibers is a well-studied route to improve the photocatalysis. Among different metal oxide semiconductors, ZnO, an n-type semiconductor in the undoped form, has proven to be an immense potential as a photocatalyst owing to its low cost, environmentally benign character, and high quantum efficiency. ZnO structures with the merit of controllable growth into nanoparticles, spindles, nanorods, and flower-like structures, show promises in photocatalytic dye decomposition. However, the nature of the powder form of ZnO makes the recycling and recovery process an arduous task; the issue can be addressed by immobilizing ZnO to nanofibrous membranes. The processes involving electrospinning and heat treatment were straightforward and delivered an outstanding performance [25, 27]. Besides, due to the wide bandgap of 3.37 eV, the photocatalytic activity of ZnO can only be triggered by UV light. Doping with metals, nonmetals, or other semiconductors can affect the ZnO bandgap, resulting in altered photocatalytic performance. Carbon-doped ZnO nanofibers lowered the bandgap energy of ZnO, which enabled the generation of oxygen and hydroxyl radicals to decompose MB under solar light excitation [46]. The stability of ZnO in mediums with different pH is also a hindrance to commercial purposes. Coating with inert oxides, such as TiO_2 and SiO_2, could show higher photostability and better photolysis due to the passivation of lattice oxygen [47]. In this case, the coating demonstrated remarkably enhanced stability in alkaline and acidic environments as a protective layer.

4.2 TiO_2 composite nanofibers

TiO_2 is one of the most studied semiconductor materials due to many advantages, including the cost-effectiveness, photocatalytic activity, biocompatibility, nontoxicity, and high stability. It has different forms, such as rutile, brookite, and anatase. The bandgaps of TiO_2 are 3.03 and 3.2 eV for rutile and anatase, respectively, and they can be activated by photons in the near UV range ($\lambda < 387$ nm). The technique of decorating TiO_2 onto nanofibers was a well-applied one to deliver the photocatalytic degradation of organic pollutants and mitigate its drawbacks as spontaneous aggregation and the problem of recovery and recycling. TiO_2-embedded CNFs have gained lots of attention in the application of dye elimination by photocatalysis. Liang et al. demonstrated that the CNFs semi-wrapped with TiO_2 could maintain consistently high photocatalytic activities against RB after five times [48]. Besides, significant efforts have been made to dope and functionalize TiO_2 to trigger the bandgap under the visible light. Qiu et al. presented a novel method of immobilizing Mo/N-codoped TiO_2 nanorods onto carbon nanofibers via two facile steps. The composite nanofibers demonstrated superb photocatalytic activity against MB, which suggested that the doping elements exhibited positive effects on dye degradation. H^+ was believed to be the main active species in the photodecomposition confirmed by trapping active species experiments [49]. The doping with other semiconductors has also demonstrated the enhancement in photocatalytic efficiency. Magnetic $ZnFe_2O_4$ with a small bandgap of 1.9 eV was successfully integrated into TiO_2 nanofibers by hydrothermal technique; the composite nanofibers promote the photoresponse under a broader region of solar light than TiO_2 [50].

4.3 Iron-based nanofibrous photocatalysts

Ion-based materials with the unique characteristic of strong magnetic response, leading to unprecedented sorption capacity and photocatalytic activities, have shown great promises in water treatment. The sizes and shapes present significant influences over the magnetic properties of iron oxide nanoparticles due to the changes in magnetic anisotropy. Among magnetic materials, FeO (wustite), Fe_3O_4 (magnetite), α-Fe_2O_3 (hematite), β-Fe_2O_3 (beta phase), γ-Fe_2O_3 (magnetite), and spinel ferrites (MFe_2O_4) have been focused on for the multiple applications including catalysis, sensors, and magnetic data storage. α-Fe_2O_3 presents weak ferromagnetism (saturation magnetization is less than 1 emu g^{-1}) at room temperature in contrast to γ-Fe_2O_3 and Fe_3O_4 (up to 92 emu g^{-1}). Thus, Fe_3O_4 and γ-Fe_2O_3 have been employed extensively to regenerate photocatalysts owing to good magnetic separation [51]. The convenience of separation by using an external magnetic field helps replace the tedious task of filtration and centrifugation for photocatalyst recovery. One prominent advantage of iron oxides is the relative narrow bandgap for the use of visible light activity, which is between 1.9 and 2.5 eV. In comparison to anatase TiO_2 (3.03–3.2 eV), which can only harvest light at a wavelength of 387 nm or below in the UV region, iron oxide-based photocatalysts prove to be superior in visible light range. The use of heterogeneous photocatalysts can accelerate the photocatalytic performance of iron oxides as a result of the enhanced visible light activation, better separation of electron-hole pair, and interfacial charge transfer. Bi_2MoO_6, which possesses a small bandgap (2.5–2.8 eV), was prepared by electrospinning; then the solvothermal method was followed to prepare 1D α-Fe_2O_3/Bi_2MoO_6 composite nanofibers [52]. The composite was demonstrated to exhibit enhanced photocatalysis in MB and RB degradation under sunlight irradiation because of the charge separation character of heterogeneous α-Fe_2O_3 and Bi_2MoO_6 composite nanomaterials.

4.4 Other photocatalysts

Different photocatalysts such as WO_3, PdO, ZrO_2, and SnO_2 have exhibited distinctive photocatalytic effects against organic dye molecules with various advantageous features such as cost-effectiveness, environmental compatibility, wide applied pH ranges, and flexible nanostructure [26, 53]. WO_3, with its bandgap varied from 2.4 to 2.8 eV, an n-type semiconductor photocatalyst, is considered as a potential photocatalyst; however, due to the fast recombination of electron and hole pairs, the photocatalytic activities of WO_3 were relatively weak. To intercept the recombination as a result of the short diffusion length of charge carriers and enhance the photocatalysis, Ma et al. introduced the grafting of Cu species by impregnation method for interfacial charge transfer effect applied in RB degradation under visible light irradiation [54]. The p- and n-type heterostructured semiconductors show better charge transfer in accordance with Fermi level equilibrium. The redistribution of charges between n-type and p-type produces inner electric fields, which facilitate the transportation of charge carriers and restrict the recombination, thus enhancing the photocatalysis. $CuCrO_2$-decorated SnO_2 composite nanofibers were synthesized by electrospinning, followed by a drop-casting method. The composite nanofibers displayed 41% better rate of constant value in comparison with pure SnO_2 [55]. Zr is in the same group IVB of elements as Ti, but ZrO_2 can only absorb 4% of solar light because of the high energy bandgap and low specific area. Lots of efforts have been made to dope ZrO_2 with other nonmetals, metals, and semiconductors in order to improve light response. The effects of different compositions of TiO_2/ZrO_2 nanofibers were reported in the photocatalytic degradation of MB dye; the nanofibers containing 40 wt% ZrO_2 displayed the best performance under visible light [56]. **Table 2** lists the photocatalytic degradation of varied metal oxide-based composite nanofibers.

No.	Photocatalyst	Dye	Light source	Time (h)	Degradation efficiency, %	Reference
1.	C-doped ZnO nanofiber	MB	Simulated solar light	0.5	> 95	[46]
2.	PdO/WO_3 NFs	MB	Visible light	24	86.4	[9]
3.	Ag-ZnO photocatalyst anchored on carbon nanofibers	MB	UV Visible light	1 2	95 95	[25]
4.	Mo/N-doped TiO_2 nanorods@CNFs	MB	Visible light	3	79.8	[49]
5	$CuCrO_2$-decorated SnO_2 composite nanofibers	MB	UV/visible light	1.5	97	[55]
6.	TiO_2/ZrO_2 composite nanofibers	MB	Visible light	3	82.7	[56]
7.	TiO_2-decorated carbon nanofibers	MB	UV	3	97.4	[30]
8.	TiO_2@carbon flexible fiber	MB	UV	18	76.06	[31]
9.	$ZnFe_2O_4$@ TiO_2 composite nanofibers	MB	Solar light	40 min	> 80	[50]
10.	α-Fe_2O_3/Bi_2MoO_6 composite nanofibers	MB RB	Sunlight	4	94.8 66.8	[52]
11.	Semi-wrapped TiO_2@carbon nanofibers	RB	UV	1	98.2	[48]
12.	WO_3/Cu (II) nanofibers	RB	Visible light	3	85	[54]

Table 2.
Comparison of different photocatalytic materials incorporated into electrospun nanofibers for dye degradation.

5. Conclusion

Electrospun composite nanofibers are advantageous in adsorbing and degrading dyestuffs with better results than using sole absorbents and promote the convenient regeneration. Many transitional metal oxides have shown efficient dye removal effects by both adsorption and photocatalytic degradation. Zeolite, graphene, GO, and MOFs have also demonstrated the high capability for dye adsorption. The mechanisms were driven by physisorption, chemisorption, and so on, which have been discussed thoroughly in this chapter. Future research should be concentrated on combining different adsorbents in the nanofibrous membranes to overcome drawbacks of each adsorbent and create hybrid nanocomposite materials with novelty and super adsorption performance. Lots of advancements are still needed to overcome the remaining issues of recyclability, secondary pollutants, and the viability in the industrial scale for the application in real dye effluents.

Author details

Duy-Nam Phan and Ick-Soo Kim*
Division of Frontier Fibers, Nano Fusion Technology Research Group, Institute for Fiber Engineering (IFES), Interdisciplinary Cluster for Cutting Edge Research (ICCER), Shinshu University, Ueda, Nagano, Japan

*Address all correspondence to: kim@shinshu-u.ac.jp

References

[1] Qureshi UA, Khatri Z, Ahmed F, Ibupoto AS, Khatri M, Mahar FK, et al. Highly efficient and robust electrospun nanofibers for selective removal of acid dye. Journal of Molecular Liquids. 2017;**244**:478-488

[2] Gopiraman M, Bang H, Yuan G, Yin C, Song K-H, Lee JS, et al. Noble metal/functionalized cellulose nanofiber composites for catalytic applications. Carbohydrate Polymers. 2015;**132**:554-564

[3] Khatri M, Ahmed F, Shaikh I, Phan D-N, Khan Q, Khatri Z, et al. Dyeing and characterization of regenerated cellulose nanofibers with vat dyes. Carbohydrate Polymers. 2017;**174**:443-449

[4] Phan D-N, Lee H, Choi D, Kang C-Y, Im SS, Kim IS. Fabrication of two polyester nanofiber types containing the biobased monomer isosorbide: Poly (ethylene glycol 1,4-cyclohexane dimethylene isosorbide terephthalate) and poly (1,4-cyclohexane dimethylene isosorbide terephthalate). Nanomaterials. 2018;**8**(2):56

[5] Abramian L, El-Rassy H. Adsorption kinetics and thermodynamics of azo-dye orange II onto highly porous titania aerogel. Chemical Engineering Journal. 2009;**150**(2):403-410

[6] Li LH, Xiao J, Liu P, Yang GW. Super adsorption capability from amorphousization of metal oxide nanoparticles for dye removal. Scientific Reports. 2015;**5**(1):9028

[7] Orozco J, Mercante LA, Pol R, Merkoçi A. Graphene-based Janus micromotors for the dynamic removal of pollutants. Journal of Materials Chemistry A. 2016;**4**(9):3371-3378

[8] Li T, Liu L, Zhang Z, Han Z. Preparation of nanofibrous metal-organic framework filter for rapid adsorption and selective separation of cationic dye from aqueous solution. Separation and Purification Technology. 2020;**237**:116360

[9] Lee H, Kim M, Sohn D, Kim SH, Oh S-G, Im SS, et al. Electrospun tungsten trioxide nanofibers decorated with palladium oxide nanoparticles exhibiting enhanced photocatalytic activity. RSC Advances. 2017;**7**(10):6108-6113

[10] Khatri Z, Nakashima R, Mayakrishnan G, Lee K-H, Park Y-H, Wei K, et al. Preparation and characterization of electrospun poly(ε-caprolactone)-poly(l-lactic acid) nanofiber tubes. Journal of Materials Science. 2013;**48**(10):3659-3664

[11] Lee H, Nishino M, Sohn D, Lee JS, Kim IS. Control of the morphology of cellulose acetate nanofibers via electrospinning. Cellulose. 2018;**25**(5):2829-2837

[12] Wei K, Kim K-O, Song K-H, Kang C-Y, Lee JS, Gopiraman M, et al. Nitrogen- and oxygen-containing porous ultrafine carbon nanofiber: A highly flexible electrode material for supercapacitor. Journal of Materials Science and Technology. 2017;**33**(5):424-431

[13] Hamano F, Seki H, Ke M, Gopiraman M, Lim CT, Kim IS. Cellulose acetate nanofiber mat with honeycomb-like surface structure. Materials Letters. 2016;**169**:33-36

[14] Kharaghani D, Tajbakhsh Z, Nam PD, Kim IS. Application of Nanowires for Retinal Regeneration. London: IntechOpen; 2019. DOI: 10.5772/intechopen.90149

[15] Zhang K, Li Z, Kang W, Deng N, Yan J, Ju J, et al. Preparation and characterization of tree-like cellulose

nanofiber membranes via the electrospinning method. Carbohydrate Polymers. 2018;**183**:62-69

[16] Zhu Q, Tang X, Feng S, Zhong Z, Yao J, Yao Z. ZIF-8@SiO2 composite nanofiber membrane with bioinspired spider web-like structure for efficient air pollution control. Journal of Membrane Science. 2019;**581**:252-261

[17] Xing R, Wang W, Jiao T, Ma K, Zhang Q, Hong W, et al. Bioinspired polydopamine sheathed nanofibers containing carboxylate graphene oxide nanosheet for high-efficient dyes scavenger. ACS Sustainable Chemistry & Engineering. 2017;**5**(6):4948-4956

[18] Phan D-N, Lee H, Huang B, Mukai Y, Kim I-S. Fabrication of electrospun chitosan/cellulose nanofibers having adsorption property with enhanced mechanical property. Cellulose. 2019;**26**(3):1781-1793

[19] Qureshi UA, Khatri Z, Ahmed F, Khatri M, Kim I-S. Electrospun zein nanofiber as a green and recyclable adsorbent for the removal of reactive black 5 from the aqueous phase. ACS Sustainable Chemistry & Engineering. 2017;**5**(5):4340-4351

[20] Ibupoto AS, Qureshi UA, Ahmed F, Khatri Z, Khatri M, Maqsood M, et al. Reusable carbon nanofibers for efficient removal of methylene blue from aqueous solution. Chemical Engineering Research and Design. 2018;**136**:744-752

[21] Gopakumar DA, Pasquini D, Henrique MA, de Morais LC, Grohens Y, Thomas S. Meldrum's acid modified cellulose nanofiber-based polyvinylidene fluoride microfiltration membrane for dye water treatment and nanoparticle removal. ACS Sustainable Chemistry & Engineering. 2017;**5**(2):2026-2033

[22] Dotto GL, Santos JMN, Tanabe EH, Bertuol DA, Foletto EL, Lima EC, et al. Chitosan/polyamide nanofibers prepared by Forcespinning® technology: A new adsorbent to remove anionic dyes from aqueous solutions. Journal of Cleaner Production. 2017;**144**:120-129

[23] Li Z, Zhou G, Dai H, Yang M, Fu Y, Ying Y, et al. Biomineralization-mimetic preparation of hybrid membranes with ultra-high loading of pristine metal–organic frameworks grown on silk nanofibers for hazard collection in water. Journal of Materials Chemistry A. 2018;**6**(8):3402-3413

[24] Gopiraman M, Kim IS. Preparation, characterization, and applications of electrospun carbon Nanofibers and its composites. In: Electrospinning and Electrospraying-Techniques and Applications. London: IntechOpen; 2019. DOI: 10.5772/intechopen.88317

[25] Pant B, Park M, Kim H-Y, Park S-J. Ag-ZnO photocatalyst anchored on carbon nanofibers: Synthesis, characterization, and photocatalytic activities. Synthetic Metals. 2016;**220**:533-537

[26] Gusain R, Gupta K, Joshi P, Khatri OP. Adsorptive removal and photocatalytic degradation of organic pollutants using metal oxides and their composites: A comprehensive review. Advances in Colloid and Interface Science. 2019;**272**:102009

[27] Malwal D, Gopinath P. Efficient adsorption and antibacterial properties of electrospun CuO-ZnO composite nanofibers for water remediation. Journal of Hazardous Materials. 2017;**321**:611-621

[28] Karunanayake AG, Navarathna CM, Gunatilake SR, Crowley M, Anderson R, Mohan D, et al. Fe_3O_4 nanoparticles dispersed on douglas fir biochar for phosphate sorption. ACS Applied Nano Materials. 2019;**2**(6):3467-3479

[29] Zhang C, Li Y, Wang F, Yu Z, Wei J, Yang Z, et al. Performance of magnetic zirconium-iron oxide nanoparticle in the removal of phosphate from aqueous solution. Applied Surface Science. 2017;**396**:1783-1792

[30] Wang H, Huang X, Li W, Gao J, Xue H, Li RKY, et al. TiO_2 nanoparticle decorated carbon nanofibers for removal of organic dyes. Colloids and Surfaces A: Physicochemical and Engineering Aspects. 2018;**549**:205-211

[31] Xiao F, Guo X, Li J, Sun H, Zhang H, Wang W. Electrospinning preparation and dye adsorption capacity of TiO_2@ carbon flexible fiber. Ceramics International. 2019;**45**(9):11856-11860

[32] Barhoum A, Pal K, Rahier H, Uludag H, Kim IS, Bechelany M. Nanofibers as new-generation materials: From spinning and nano-spinning fabrication techniques to emerging applications. Applied Materials Today. 2019;**17**:1-35

[33] Habiba U, Siddique TA, Li Lee JJ, Joo TC, Ang BC, Afifi AM. Adsorption study of methyl orange by chitosan/ polyvinyl alcohol/zeolite electrospun composite nanofibrous membrane. Carbohydrate Polymers. 2018;**191**:79-85

[34] Lee JJL, Ang BC, Andriyana A, Shariful MI, Amalina MA. Fabrication of PMMA/zeolite nanofibrous membrane through electrospinning and its adsorption behavior. Journal of Applied Polymer Science. 2016;**134**(6):44450. DOI: 10.1002/app.44450

[35] Zhan Y, Guan X, Ren E, Lin S, Lan J. Fabrication of zeolitic imidazolate framework-8 functional polyacrylonitrile nanofibrous mats for dye removal. Journal of Polymer Research. 2019;**26**(6):145

[36] Mercante LA, Facure MHM, Locilento DA, Sanfelice RC, Migliorini FL, Mattoso LHC, et al. Solution blow spun PMMA nanofibers wrapped with reduced graphene oxide as an efficient dye adsorbent. New Journal of Chemistry. 2017;**41**(17):9087-9094

[37] Ma F-f, Zhang D, Huang T, Zhang N, Wang Y. Ultrasonication-assisted deposition of graphene oxide on electrospun poly(vinylidene fluoride) membrane and the adsorption behavior. Chemical Engineering Journal. 2019;**358**:1065-1073

[38] Tran HN, Wang Y-F, You S-J, Chao H-P. Insights into the mechanism of cationic dye adsorption on activated charcoal: The importance of π–π interactions. Process Safety and Environmental Protection. 2017;**107**:168-180

[39] Mahmoodi NM, Oveisi M, Taghizadeh A, Taghizadeh M. Synthesis of pearl necklace-like ZIF-8@chitosan/ PVA nanofiber with synergistic effect for recycling aqueous dye removal. Carbohydrate Polymers. 2020;**227**:115364

[40] Jia S, Tang D, Peng J, Sun Z, Yang X. β-Cyclodextrin modified electrospinning fibers with good regeneration for efficient temperature-enhanced adsorption of crystal violet. Carbohydrate Polymers. 2019;**208**:486-494

[41] Zhang J, Yan Z, Ouyang J, Yang H, Chen D. Highly dispersed sepiolite-based organic modified nanofibers for enhanced adsorption of Congo red. Applied Clay Science. 2018;**157**:76-85

[42] Hosseini SA, Vossoughi M, Mahmoodi NM, Sadrzadeh M. Clay-based electrospun nanofibrous membranes for colored wastewater treatment. Applied Clay Science. 2019;**168**:77-86

[43] Ma Y, Qi P, Ju J, Wang Q, Hao L, Wang R, et al. Gelatin/alginate

composite nanofiber membranes for effective and even adsorption of cationic dyes. Composites Part B: Engineering. 2019;**162**:671-677

[44] Liang Z, Zhao Z, Sun T, Shi W, Cui F. Enhanced adsorption of the cationic dyes in the spherical CuO/meso-silica nano composite and impact of solution chemistry. Journal of Colloid and Interface Science. 2017;**485**:192-200

[45] Yazdi MG, Ivanic M, Mohamed A, Uheida A. Surface modified composite nanofibers for the removal of indigo carmine dye from polluted water. RSC Advances. 2018;**8**(43):24588-24598

[46] Gadisa BT, Appiah-Ntiamoah R, Kim H. In-situ derived hierarchical ZnO/Zn-C nanofiber with high photocatalytic activity and recyclability under solar light. Applied Surface Science. 2019;**491**:350-359

[47] Wang Y, Zheng Y-Z, Lu S, Tao X, Che Y, Chen J-F. Visible-light-responsive TiO_2-coated ZnO:I nanorod array films with enhanced photoelectrochemical and photocatalytic performance. ACS Applied Materials & Interfaces. 2015;**7**(11):6093-6101

[48] Liang Y, Zhou B, Li N, Liu L, Xu Z, Li F, et al. Enhanced dye photocatalysis and recycling abilities of semi-wrapped TiO_2@carbon nanofibers formed via foaming agent driving. Ceramics International. 2018;**44**(2):1711-1718

[49] Qiu J, Liu F, Yue C, Ling C, Li A. A recyclable nanosheet of Mo/N-doped TiO_2 nanorods decorated on carbon nanofibers for organic pollutants degradation under simulated sunlight irradiation. Chemosphere. 2019;**215**:280-293

[50] Al-Meer S, Ghouri ZK, Elsaid K, Easa A, Al-Qahtani MT, Shaheer Akhtar M. Engineering of magnetically separable $ZnFe_2O_4$@ TiO_2 nanofibers for dye-sensitized solar cells and removal of pollutant from water. Journal of Alloys and Compounds. 2017;**723**:477-483

[51] Wu W, Changzhong J, Roy VAL. Recent progress in magnetic iron oxide–semiconductor composite nanomaterials as promising photocatalysts. Nanoscale. 2015;**7**(1):38-58

[52] Zhao J, Lu Q, Wang Q, Ma Q. α-Fe_2O_3 nanoparticles on Bi_2MoO_6 nanofibers: One-dimensional heterostructures synergistic system with enhanced photocatalytic activity. Superlattices and Microstructures. 2016;**91**:148-157

[53] Wang X, Dou L, Yang L, Yu J, Ding B. Hierarchical structured MnO2@SiO2 nanofibrous membranes with superb flexibility and enhanced catalytic performance. Journal of Hazardous Materials. 2017;**324**:203-212

[54] Ma G, Lu J, Meng Q, Lv H, Shui L, Zhang Y, et al. Synergistic effect of Cu-ion and WO_3 nanofibers on the enhanced photocatalytic degradation of Rhodamine B and aniline solution. Applied Surface Science. 2018;**451**:306-314

[55] Dursun S, Kaya IC, Kalem V, Akyildiz H. UV/visible light active $CuCrO_2$ nanoparticle–SnO_2 nanofiber p–n heterostructured photocatalysts for photocatalytic applications. Dalton Transactions. 2018;**47**(41):14662-14678

[56] Yasin AS, Obaid M, El-Newehy MH, Al-Deyab SS, Barakat NAM. Influence of $TixZr(1-x)O_2$ nanofibers composition on the photocatalytic activity toward organic pollutants degradation and water splitting. Ceramics International. 2015;**41**(9, Part B):11876-11885

Chapter 8

Nanosilica Composite for Greenhouse Application

Malek Alghdeir, Khaled Mayya and Mohamed Dib

Abstract

This work represents the results of experiments on silicon dioxide insulation materials mixed with low-density polyethylene (LDPE) at a different proportion to prevent the transmittance of IR domain and to allow the transmittance of ultraviolet–visible (UV–Vis) domains, so we can keep the thermal radiation of the ground in the greenhouse. The mechanical properties of nanocomposites such as tensile were evaluated and discussed. Several ratios of nanosilica particles were employed to fabricate low-density polyethylene (LDPE) composites using melt mixing and hot molding methods. Six of composite films from different ratios (0.5, 1, 2.5, 5, 7.5, and 10 wt% nanosilica) were prepared. The obtained composite films were characterized and identified by ultraviolet–visible (UV–Vis) spectroscopy and Fourier transform infrared spectroscopy (FTIR). Thermal stability of samples was evaluated by thermogravimetric analysis (TGA). Surface morphology of samples was investigated by scanning electron microscopy (SEM). At specific mixing ratio, the ultraviolet–visible transmittance is allowed, while far infrared radiation transmittance was prohibited, and that will be explained in details. Optical measurements show that the composite films prevent the transmission of IR radiation near 9 μm and allow UV–Vis transmission during sun-shining time. The mechanical behavior was studied using tensile tests for nanosilica-reinforced LDPE composite. The sample with an addition of 1 wt% nanosilica has successfully enhanced the mechanical properties of LDPE material.

Keywords: silicon dioxide (SiO_2), low-density polyethylene (LDPE), composite barrier films, Fourier transform infrared spectroscopy (FTIR)

1. Introduction

Polymeric materials are widely used in greenhouses and in food packaging. Typical examples of such materials are polyethylene terephthalate (PET), polyethylene (PE), and polypropylene (PP) [1, 2].

In recent years, many studies have been done, and much effort has been devoted to polymer nanocomposites which have attracted a great attention from scientists [3]. The literature contains a plethora of experiments illustrating the thus involved parameters. At turns and among others, fabrication technique, matrix nature, nanofiller shape factor, and complex nature of the additives may be the determinants on the end-result properties of the polymer nanocomposites [4]. The dispersion of nanoparticles in the polymer matrix [5, 6] and the property of the interface between nanoparticle and polymer are regarded as key factors affecting the insulating properties of nanocomposites [7, 8]. Polymer nanocomposites often show superior mechanical properties compared to the conventional composites at a lower loading of the nanoparticles [9]. So far, a few researches have systematically studied the effects of different nanoparticles on the performances of composite materials [3].

The excellent properties of silicon dioxide film have attracted attention in industry and academia, due to its hardness, corrosion resistance, anti-resistance [10], dielectric properties [11], optical transparency, etc. [12]. Silicon dioxide as a thin film is widely used to improve the surface properties of materials. This is why silicon dioxide films are used in many fields as in antireflection coating field [13]. Silicon dioxide films are used as barrier layers in polymer packaging materials in the packaging industry. Most of the modern packaging materials do not provide an efficient barrier against the permeation of gases. This leads to drink and food not getting rotten quickly. Because of this, a silicon dioxide film deposited on the surface of polymer packaging becomes indispensable and popular. In addition, silicon dioxide films can be also used as corrosion protective layers of metals. Besides, the preparation of silica with high quality is always an important aim of scientific research because of the universal application of silicon dioxide films in various fields [14].

Currently, a number of different barrier coating technologies are being developed. Theoretically, a barrier function can be incorporated into a plastic-based packaging material via two different means, either by mixing a barrier material with the base polymer or by coating a layer of the barrier material [15, 16].

Presently, the traditional and simplest method of preparing polymer/silica composites is direct mixing of the silica into the polymer. The mixing can generally be done by melt blending and solution blending. This mixing process always results in the effective dispersion of the silica nanoparticles in the polymer matrix. The basic difficulty is when they usually tend to agglomerate [17].

This work represents the results of experiments on silicon dioxide insulation materials mixed with LDPE at a different proportion to prevent the transmittance of IR domain and to allow the transmittance of UV–Vis domains, so we can keep the thermal radiation of the ground in the greenhouse. The mechanical properties of nanocomposites such as tensile were evaluated and discussed.

Samples of neat LDPE and nanocomposites in different ratios (0.5, 1, 2.5, 5, 7.5, and 10 wt% SiO_2) were produced. Many material properties were investigated and will be discussed.

By studying blackbody thermal radiation, all objects with a temperature above absolute zero emit energy in the form of electromagnetic radiation. A blackbody is a model or theoretical body which absorbs all radiation falling

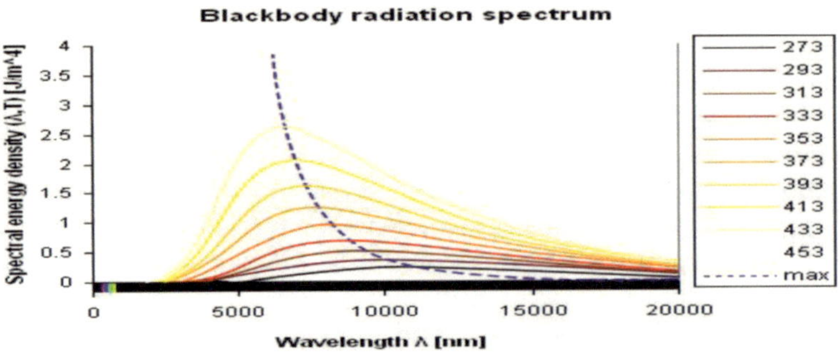

Figure 1.
Blackbody radiation spectra at 273, 293, 313, 333, 373, 393, 413, 433, and 453 K [14].

on it. It is a hypothetical object which is a "perfect" absorber and a "perfect" emitter of radiation. The electromagnetic radiation emitted by a blackbody has a specific intensity and spectrum that depend only on the body's temperature; the thermal radiation spontaneously emitted by ordinary objects, for example, plants and land, can be approximated as blackbody radiation. **Figure 1** shows the blackbody radiation spectrum at different several temperatures. We are interested in the vicinity of 10 μm (9–11 μm), because at the temperatures near 0°C (273 K), the thermal radiation from the ground is maximum at 10 μm while at the temperature 30°C (303 K), thermal radiation from the ground is maximum at 9.5 μm [14].

2. Material and methods

2.1 Materials

Low-density polyethylene (LDPE) which is a thermoplastic made from the monomer ethylene with a density of 0.922 g/cm^3 was purchased from Saudi Basic Industries Corporation (SABIC). High-purity fumed nanosilica (purity >99%) with the trademarks of A200 with an average particle size of ~12 nm was obtained from Evonik Degussa AG (Germany). **Table 1** shows the information on basic physical and chemical properties for fumed nanosilica.

2.2 Sample preparation

Samples were prepared by blending LDPE in different ratios (0.5, 1, 2.5, 5, 7.5, and 10 wt% SiO_2) and making plates from composite material. In different nanosilica ratios, nanosilica composites were mechanically mixed with LDPE granules at the processing temperature of 130°C and speed at 50 rpm min^{-1} for 10 min using the internal mixer (Brabender Plasti-Corder PL-2200, W50, Germany). Films of the neat LDPE with dimension of 10 cm × 10 cm × 120 μm and nanocomposites were prepared by a hot press method at the temperature of 140°C and the pressure of 50 bars.

Properties	Value
Physical state	Solid
Color	White
Form	Powder
pH	3.7–4.5
Melting point/range	Approx. 1700°C
Surface area	200 ± 25 m^2/g
Density	Approx. 2.2 g/cm^3
Thermal decomposition	>2000°C
Water solubility	>1 mg/l
Loss on drying	≤1.5% (2 h at 105°C)
Silica content based on ignited material	>99.8%

Table 1.
Information on basic chemical and physical properties for nanosilica.

3. Results and discussion

3.1 Infrared spectroscopic study

The transmittance of samples was examined by Fourier transform infrared (FTIR) spectroscopy (FTIR spectrometer, VERTEX 70/70v from Bruker™ Optics) in the wavelength range of 1–25 μm. **Figure 2** shows the transmittance spectra of the SiO_2/LDPE films in different ratios.

Different absorption peaks could be identified in the MIR range. The first one at ~3 μm caused by OH group and other peaks at ~9 μm, ~12 μm, and ~21 μm, due to Si—O—Si resonance mode of vibrations [18]. Some of these peaks also involve the LDPE substrate in the IR absorption spectra. The peak at 9 μm gives the SiO_2 its importance and allows it to be used in this application. We observe a decrease in transmittance when the mixing ratio of SiO_2 increases. The changes in the average transmittance for wavelengths ranging from 7 to 10.5 μm are shown in **Figure 3**. We notice a sharp decline in transmittance when the ratio of SiO_2 is increased.

3.2 Ultraviolet-visible spectroscopy study

The optical transmittance measurements of LDPE/silica nanocomposite substrate films were carried out with a UV–Vis–NIR spectrophotometer (UV Spectrophotometer A560AOE instruments) at normal incident of light in the wavelength range of 200–1100 nm. **Figure 4** shows the transmittance spectra of the samples. The UV spectra show that the composite substrates (0.5, 1, 2.5, 5, and 7.5 wt% SiO_2) have no significant effect on the transmittance. On the other hand, a significant decrease in the transmittance is observed with a mixture ratio of 10 wt% SiO_2 compared to the LDPE without mixing. This decrease is addressed in Section 3.

3.3 Mechanical property study

For mechanical property studies, the test specimens for tensile was molded and cut according to the dimensions specified in **Table 2**. The tensile test was measured using a traction compression machine Adamel Lhomargy DY34 under atmospheric conditions. Average five samples were tested and the stress-strain curves were recorded. Crosshead speed for tensile tests was carried at 5 mm/min. **Table 3** shows the summary of tensile data for the control sample and six composites of LDPE.

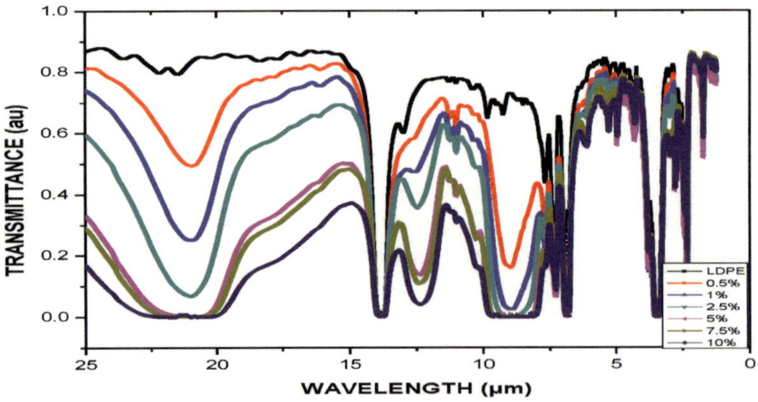

Figure 2.
FTIR spectra for different ratios of LDPE/silica nanocomposite film.

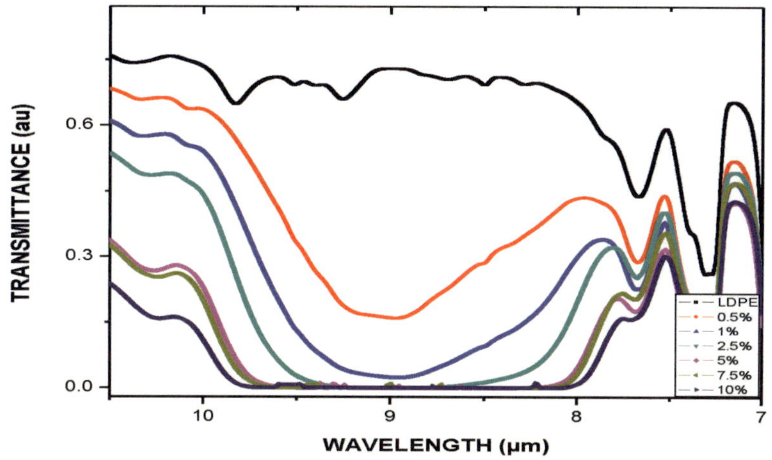

Figure 3.
FTIR spectra in the range of 7–10.5 μm.

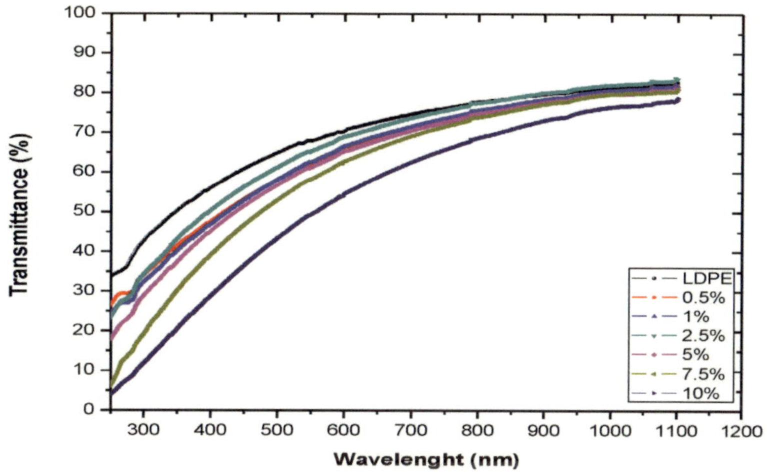

Figure 4.
UV–Vis spectra of LDPE/silica nanocomposite film.

Specifications	Dimensions (mm)
Sample length	75
Display ends	12.5 ± 1
Length of the active part	25 ± 1
Display the effective part	4 ± 0.1
External radius	8 ± 0.5
Internal radius	12.5 ± 1

Table 2.
Specified dimension of samples for tensile test.

Figure 5 shows the stress–strain curves from tensile tests for LDPE/silica nanocomposites. **Figure 5** shows that the stress at break gradually increasing with the increase of silica loading up to 1 wt%. This result suggests that the fine silica particle

Material	Property	Average	Std. dev.
LDPE control	Maximum load, N	58	4.47
	Tensile strength, MPa	8.18	0.60
	% elongation at break	250	48
	Modulus of elasticity, MPa	64.47	5.1
LDPE 0.5 wt% SiO_2	Maximum load, N	60	0
	Tensile strength, MPa	7.91	0.22
	% elongation at break	307	52
	Modulus of elasticity, MPa	95.4	9.2
LDPE 1 wt% SiO_2	Maximum load, N	60	0
	Tensile strength, MPa	8.78	0.09
	% elongation at break	374	40
	Modulus of elasticity, MPa	88.4	12
LDPE 2.5 wt% SiO_2	Maximum load, N	60	0
	Tensile strength, MPa	8.64	0.17
	% elongation at break	362	92
	Modulus of elasticity, MPa	96	3.5
LDPE 5 wt% SiO_2	Maximum load, N	60	0
	Tensile strength, MPa	8.27	0.10
	% elongation at break	231	71
	Modulus of elasticity, MPa	122	11
LDPE 7.5 wt% SiO_2	Maximum load, N	60	0
	Tensile strength, MPa	8.35	0.07
	% elongation at break	253	31
	Modulus of elasticity, MPa	112	16
LDPE 10 wt% SiO_2	Maximum load, N	60	0
	Tensile strength, MPa	8.38	0.19
	% elongation at break	118	16
	Modulus of elasticity, MPa	122	21

Table 3.
Summary of tensile data for the control sample and six composites of LDPE.

would reinforce and orient along the direction of stress and this has contributed to the increase of tensile strength of the nanocomposite with the addition of 1 wt% of nanosilica particles.

The stress–strain curves also illustrate that there was a significant increase of elongation at break values of LDPE with the incorporation of nanosilica particle into the nanocomposite. This result indicates that the incorporation of nanosilica particle would improve the interaction between the molecules. At lower weight percentage, the addition of nanosilica in LDPE matrix increases surface interaction bonding between the molecules. The nanoparticles may be trapped inside entanglements resulting in a restriction on the polymer overall chain mobility.

Figure 5 shows that both of the stress at break and elongation at break (strain) of nanocomposite achieved the highest values with 1 wt% loading of silica particles.

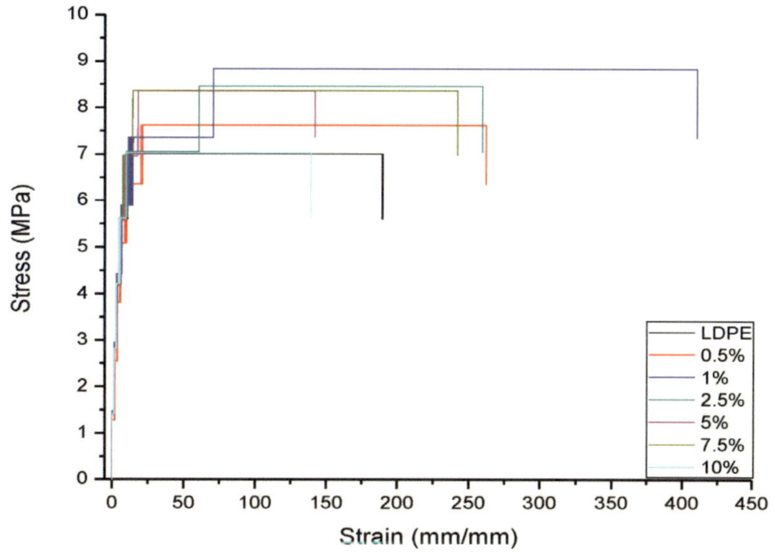

Figure 5.
Typical stress–strain curves of neat LDPE and LDPE/silica nanocomposites.

Above 1 wt% silica loading, both the stress and elongation at break showed a gradual drop. These results are attributed to the reinforcing effect of the silica particles.

A higher amount of silica particles would reduce the reinforcing effect and mechanical properties of the nanocomposites due to poor dispersion and agglomeration of silica particles. The agglomerated silica particles with larger particle size would serve as flaws and stress concentration for crack initiation, resulting in poor tensile properties.

In this study, the nanoparticles were well dispersed at lower loading (0.5 to 1 wt%) of silica particles. The reinforcing effect of this small amount of nanoparticle loadings with the huge specific surface area has a dramatically larger total interface area for reinforcement efficiency.

In **Figure 6** all nanocomposites show the same average tensile strength compared to that of LDPE. This is not unexpected as large changes in tensile strength and extensibility and cannot be expected at a loading below of 10 wt.% of nanofiller [9]. **Figure 6** illustrates the effect of nanosilica particles to Young's modulus of LDPE matrix. As shown in **Figure 6**, Young's modulus increased with the addition of silica nanoparticles. This suggests that the incorporation of silica nanoparticles would improve the stiffness of LDPE. At low silica contents of 1 wt%, the nanocomposite exhibits an interactive structure with the matrix. Strong interfacial interaction will enable the load to be transferred easily across the nanoparticle-matrix interface. This will contribute to the increase of Young's modulus and tensile strength of the nanoparticle-reinforced composites. However, the agglomeration of particles occurred with the increase of silica loadings. The high amount of nanoparticle loadings did not participate in homogeneous interactive bonding with LDPE. The weak interaction between particle and matrix has caused lower tensile properties due to the debonding of particle from matrix prior to the plastic deformation of the matrix.

3.4 Greenhouses thermal study

We built a mini greenhouse of LDPE without mixing and another of LDPE mixed with 2.5 wt% SiO_2. We also built a third mini greenhouse of silica glass window (glass thickness is 6 mm, the transmittance from 350 to 1100 nm is 88% approximately) (see **Figure 7**). All the three greenhouses are cubic with a side of 20 cm. Inside each greenhouse, we put a small plant. This plant was previously grown under similar conditions.

The temperature inside each greenhouse was measured using identical temperature sensors (Tecnologic with resolution 0.1°C). The external temperature was also measured using an identical sensor. All the measurements were made at the same moment every 30 minutes starting from 1:00 PM until 6:00 AM the next day. **Figure 8** shows the temperature variations inside the three greenhouses along with the external air temperature. An increase in the temperature inside the greenhouse mixed with 2.5 wt% SiO_2 is noticed. This increase is estimated to be more than 2°C than the LDPE greenhouse without mixing (2°C overall and 2.2°C between 11:00 PM and 5:00 AM). We also noticed that the transmittance of the greenhouse mixed with 2.5 wt% SiO_2 approaches that of the glass house very much (see the green and blue triangles in **Figure 8**). In fact, the average temperature difference is about 0.14°C overall, and the two temperatures between 11:00 PM and at 5:00 AM match to each other very well.

By studying the IR transmission in **Figures 2**, **3**, and **9**, a decrease in the transmittance near 9 μm with increasing mixture ratios is noticed. This result explains the rise in temperature inside the mini greenhouses (shown in **Figure 9**). The LDPE/silica nanocomposite barrier films preserve the thermal radiation of the ground. Thus, the internal temperature inside the greenhouse is maintained.

One can also notice that in the vicinity of 9 μm, the transmittance of the sample with a ratio of 5 wt% SiO_2 is very close to that with a ratio of 2.5 wt% SiO_2. We deduce that it may not be very beneficial to go beyond a ratio of 2.5 wt% SiO_2.

By studying the UV–Vis transmission in **Figure 4**, a significant decrease is noticed in the transmittance of the film with a ratio of 10 wt% SiO_2, compared with the other films of less ratios (0.5, 1, 2.5, 5, and 7.5 wt% SiO_2). These five composite barrier films do not have any significant effect on the transmittance compared with that of the LDPE without mixing. Thus, the film with a ratio of 2.5 wt% SiO_2 composite film was adopted to build the mini greenhouse. It has no effect on the UV–Vis transmission but it reduces a maximum transmission of the IR radiation around 9 μm.

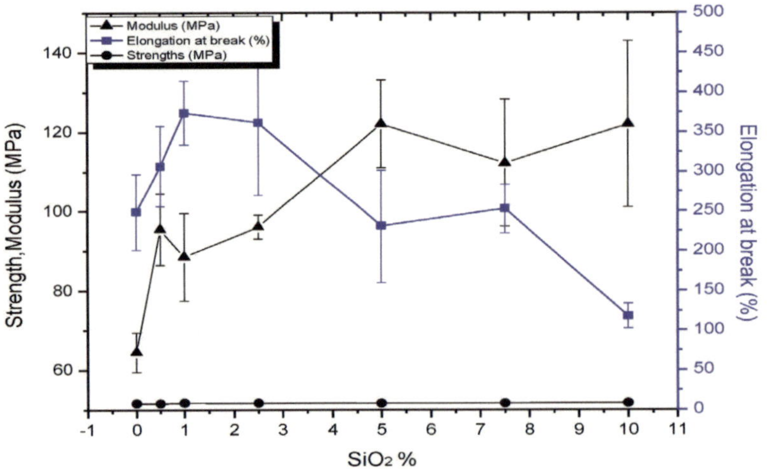

Figure 6.
The tensile strength, elongation at break and Modulus of elasticity as a function of different ratios of SiO_2.

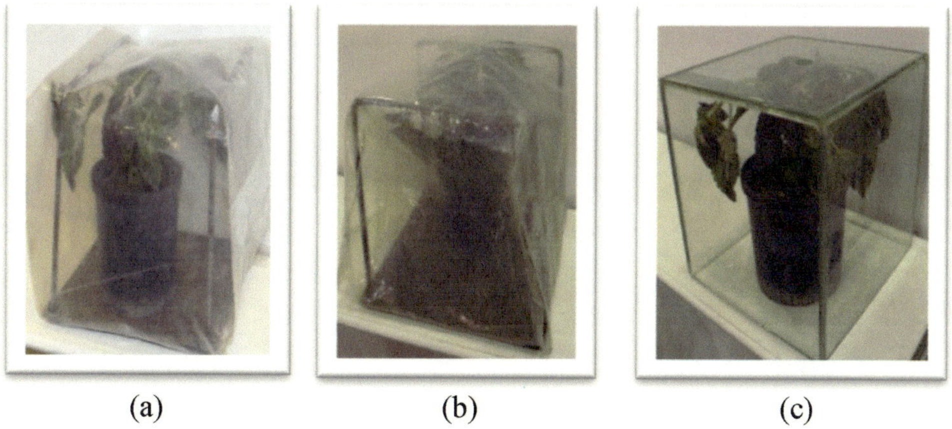

Figure 7.
The three greenhouses. (a) LDPE without mixing, (b) LDPE mixing with 2.5 wt% SiO_2, and (c) silica glass.

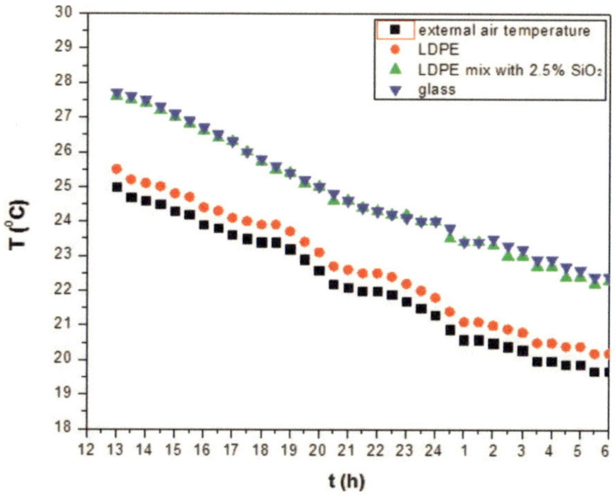

Figure 8.
The variations of difference temperature (ΔT), between the temperature inside the greenhouse and the temperature in the external air, during the time starting from 1 P.M. (13) until 6 A.M.

The refractive index of LDPE in the visible domain is 1.51, while the imaginary part is $k = 0$ [19]. It is very close to the real part value of the refractive index of SiO_2 which is equal to 1.43 [20]. Therefore, there should not be any significant change in the transmittance of the LDPE, in the visible range, when mixed with SiO_2. This is clearly seen in **Figure 4** except for the last case where the ratio of the SiO_2 is 10 wt%. Consequently, there should not be any significant change in the greenhouse temperature during sun-shining time. The significant reduction in the transmittance in the case where the ratio of SiO_2 is 10 wt% is probably due to Mie scattering [21–23].

3.5 SEM study

Figure 10 shows the SEM images at a magnification of ×2000 that show no obvious signs of agglomeration of the filler. At this magnification, significant agglomeration should be discernible. While some idea of the level of dispersion might be ascertained by scanning electron microscopy (SEM), the small areas

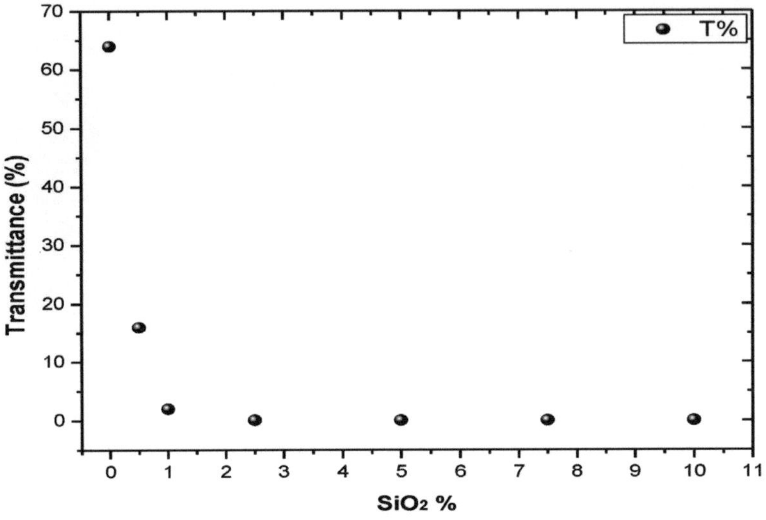

Figure 9.
Transmittance at 9 μm as a function of different ratios of SiO_2.

Figure 10.
Scanning electron micrograph of nanocomposite LDPE 2.5 wt% SiO_2 *sample at ×2000 magnification.*

sampled by this technique do not yield an average picture of the sample. Also, transmission electron microscopy (TEM) is better suited than SEM for studying nanoparticle dispersion [9].

3.6 Thermal analysis

The decomposition temperatures (Td) for LDPE and LDPE nanocomposite were obtained by thermogravimetric analysis (TGA). Samples (10 mg) placed in

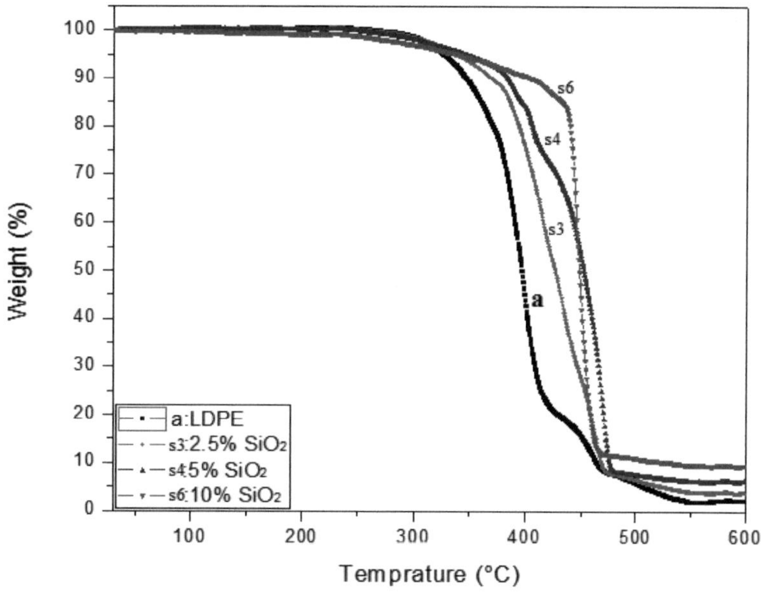

Figure 11.
TGA curves of neat LDPE and LDPE/SiO_2 nanocomposites under nitrogen flow.

an alumina holder were submitted to a heating cycle at a rate of 10°C/min starting from room temperature (25°C) up to 600°C, under nitrogen atmosphere.

Figure 11 shows the decomposition profile (percentage of weight loss with temperature) for LDPE and LDPE/silica nanocomposites performed in one step. The presence of SiO_2 in the nanocomposite confers thermal stability to it shown by the continuous shifting of the curve to higher temperatures as the SiO_2 content increased. These results may be attributed to the silica layers acting as an insulator and a barrier to mass transport from volatile substances generated during the decomposition [24].

4. Conclusion

Throughout this study, silicon dioxide films with different ratios (0.5, 1, 2.5, 5, 7.5, and 10 wt% SiO_2) were mixed with low-density polyethylene (LDPE) polymer, using melt mixing technique.

The effect of incorporation of 0.5–10 wt% of silica particles to a tensile property of LDPE matrix was investigated. The results showed that the addition of 1 wt% of nanosilica has successfully enhanced the tensile and elongation at break of the nanosilica-filled LDPE material. The incorporation of >1 wt% of nanosilica particles had caused agglomeration and uneven distribution of the particles throughout the LDPE matrix.

These LDPE/silica nanocomposites were used to build a mini greenhouse. SiO_2 reduces the transmission of radiation near 9 μm and allows the transmission of the ultraviolet and visible radiations to pass through them during daytime (period of sunshine, without being exposed to direct sunshine.) Thus, we were able to preserve the thermal radiation of the ground by raising the internal temperature of the greenhouse up to more than 2°C than that of the same greenhouse without mixing. The temperature inside the LDPE/silica greenhouse was found to be almost identical to that inside the glass greenhouse.

Statistically speaking the conclusions are acceptable because the experiment was replicated many time. The main gain is the fact that the LDPE/silica greenhouse has the same temperature as the glass-made greenhouse.

Author details

Malek Alghdeir*, Khaled Mayya and Mohamed Dib
Applied Physics Department, Higher Institute for Applied Sciences and Technology, Damascus, Syria

*Address all correspondence to: malekghdeir@yahoo.com

References

[1] Kuzminova A, Shelemin A, Petr M, Kylian O, Biederman H. Barrier coatings on polymeric foils for food packaging. In: WDS'13 Proceedings of Contributed Papers, Part III. 2013. pp. 128-133

[2] Kylián O, Choukourov A, Hanyková L, Biederman H. Plasma technology for polymer food packaging materials. In: Ecosustainable Polymer Nanomaterials for Food Packaging: Innovative Solutions, Characterization Needs, Safety and Environmental Issues. CRC Press; 2013. p. 119

[3] Ju S, Chen M, Zhang H, Zhang Z. Dielectric properties of nanosilica/low-density polyethylene composites: The surface chemistry of nanoparticles and deep traps induced by nanoparticles. eXPRESS Polymer Letters. 2014;**8**(9):682-691

[4] Fréchette MF, Ghafarizadeh SB, Ahn TT, Vadeboncoeur S, Guo M, DavidE.LDPEnanocompositescontaining functionalized SiO_2 molecular structures: Properties associated with a ball-milled preparation. In: 2017 1st International Conference on Electrical Materials and Power Equipment (ICEMPE). IEEE. 2017. pp. 207-210

[5] Yin Y, Chen J, Yang J, Xiao D, Tu D, Yin R, et al. Effect of space charge in nanocomposite of LDPE/TiO_2. In: Proceedings of the 7th International Conference on Properties and Applications of Dielectric Materials. Nagoya, Japan, Vol. 3. 2003. pp. 913-916. DOI: 10.1109/ICPADM.2003.1218570

[6] Yang J-M, Wang X, Zheng C-J, Zhao H, Lei Q-Q. DC conduction properties of SiO_2/LDPE nanocomposite. In: 10th IEEE International Conference on the Properties and Applications of Dielectric Materials (ICPADM), Bangalore, India. 2012. pp. 1-4. DOI: 10.1109/ICPADM.2012.6318950

[7] Wang W, Li S, Tang F, Li J. Characteristics on breakdown performance of polyethylene/silica dioxide nano-composites. In: 2012 Annual Report Conference on Electrical Insulation and Dielectric Phenomena. Montreal, Canada. 2012. pp. 521-524. DOI: 10.1109/CEIDP.2012.6378834

[8] Schmidt D, Shah D, Giannelis EP. New advances in polymer/layered silicate nanocomposites. Current Opinion in Solid State and Materials Science. 2002;**6**:205-212. DOI: 10.1016/s1359-0286(02)00049-9

[9] Redhwi HH, Siddiqui MN, Andrady AL, Hussain S. Durability of LDPE nanocomposites with clay, silica, and zinc oxide—Part I: Mechanical properties of the nanocomposite materials. Journal of Nanomaterials. 2013;**2013**:21

[10] Pal A. Low-Power VLSI Circuits and Systems. India: Springer; 2015. pp. 19-42. DOI: 10.1007/978-81-322-1937-8

[11] Yasunas A, Kotov D, Shiripov V, Radzionay U. Low-temperature deposition of silicon dioxide films in high-density plasma. Semiconductor Physics Quantum Electronics & Optoelectronics. 2013;**16**(2):216-219

[12] Delimi A, Coffinier Y, Talhi B, Boukherroub R, Szunerits S. Investigation of the corrosion protection of SiOx-like oxide films deposited by plasma-enhanced chemical vapor deposition onto carbon steel. Electrochimica Acta. 2010;**55**(28):8921-8927. DOI: 10.1016/j.electacta.2010.08.008

[13] Li WT, Boswell R, Samoc M, Samoc A, Wang RP. Thin Solid Films. 2008;**516**:5474-5477

[14] Alghdeir M, Mayya K, Dib M, Alghoraibi I. Characterization of SiO_2/LDPE composite barrier films. Journal of Materials and Environmental Science. 2018;**9**(7):2042-2050

[15] Liu Z, Sun Z, Ma X, Yang CL. Characterization of composite SiOx/polymer barrier films. Packaging Technology and Science. 2013;**26**:70-79

[16] Lange J, Wyser Y. Recent innovations in barrier technologies for plastic packaging—A review. Packaging Technology and Science: An International Journal. 2003;**16**(4):149-158

[17] Zou H, Wu S, Shen J. Polymer/silica nanocomposites: Preparation, characterization, properties, and applications. Chemical Reviews. 2008;**108**(9):3893-3957

[18] Kitamura R, Pilon L, Jonasz M. Optical constants of silica glass from extreme ultraviolet to far infrared at near room temperature. Applied Optics. 2007;**46**(33):8118-8133

[19] Mitchell BS. An Introduction to Materials Engineering and Science for Chemical and Materials Engineers. Hoboken, New Jersey: John Wiley & Sons, Inc.; All rights reserved. Published simultaneously in Canada; 2004. p. 901

[20] Xiong C, Xu W, Zhao Y, Xiao J, Zhu X. New design graded refractive index antireflection coatings for silicon solar cells. Modern Physics Letters B. 2017;**31**(19-21):1740028

[21] Iqbal M. An Introduction to Solar Radiation. Canada: Academic Press, Ontario: Elsevier; 2012. p. 116

[22] Alghdeir M, Mayya K, Dib M. Characterization of nanosilica/low-density polyethylene nanocomposite materials. Journal of Nanomaterials. 2019. p. 7

[23] Alghdeir M, Mayya K, Dib M, Alghoraibi I. Development of optical barrier films on flexible polymer substrates. Moroccan Journal of Chemistry. 2019;**7**(2):354-362

[24] Pavlidou SPC. A review on polymer—layered silicate nanocomposites. Progress in Polymer Science. 2008;**33**:1119-1198

Section 4

Analysis and Performance of Composites

Chapter 9

Damage Identification and Assessment Using Lamb Wave Propagation Parameters and Material Damping in FRP Composite Laminates

Beera Satish Ben and Beera Avinash Ben

Abstract

A methodology for identify damage in the fiber reinforced polymer (FRP) composite has been proposed in this article. The Lamb wave dispersion theory was used to find the fitted peak frequency and loss less finite element model was used to find the modal frequencies in the composite laminates. The change in modal parameters with respect to undamaged and damaged specimen has been considered for the structural diagnosis. The combined finite element and Lamb wave method has been used to obtain damping parameters. The damping capacity was calculated at higher frequency and smaller amplitudes by using hybrid method. The Lamb waves were generated using ultrasonic pulse generator setup. The proposed method was implemented on FRP laminates (CFRP and GFRP) and the results were compared with bandwidth method.

Keywords: carbon fibers, Lamb wave, glass fibers, vibration, damping, acoustic emission

1. Introduction

Composite materials with advanced properties like, high specific strength and fatigue resistance are being used for many components of aircraft structures in recent era. However, they have high chances of failure when they are subjected to low velocity-impact, which could lead to barely visible impact damage (BVID). BVID are considered to be internal defects which can lead to catastrophic accidents in service. Damage is defined as the changes introduced into a system that leads to affect adversely to its current or future performance. Damage assessment techniques in the structural dynamics have been divided into three categories such as linear, non-linear and transient vibrational measurements. The change in modal parameters such as natural frequencies and material damping can be considered as the prevalent damage detection methods in structural assessment procedure. The existing damage in a structure, leads to the reduction in stiffness and consequently decreasing of the natural frequencies of the system.

According to Doebling et al. [1] damage detection by using vibration measurement in elastomers has been first reported by Lifshitz and Rotem [2]. They used the changes in the dynamic moduli and change in the natural frequencies to detect damage.

Many researchers identified crack depth and location from the dependency of the first two structural Eigen frequencies and presented contour graph. The superposed contour of the frequencies variations between the undamaged and damaged structures is used to identification the damage. The intersection point of superposed contour allows identifying both the crack depth and location [3].

According to Gillich and Praisach [4] the change in damping and the friction between crack surfaces lead to dissipative effects. The advantages of using changes in damping is that the cracks allows changes in natural frequencies due to uncertainties and cause important changes in the damping factor allowing damage detection.

Kyriazoglou and Guild [5] predicted damping parameters of GFRP and CFRP laminates by using finite element model. Most of the damping related calculations and experiments were carried out by Berthelot and Sefrani [6–8] for various composites using Ritz Method. They have performed damping analysis of composite plate and structures by using this method [9–11].

Chen and Gibson [12] have studied damping mechanisms in composites which involves a variety of energy dissipation mechanisms. The vibrational parameters such as frequency and amplitude in fiber-reinforced polymers that depend on energy dissipation mechanisms are studied with nondestructive evaluation.

Many researchers have used nondestructive evaluation (NDE) techniques to characterize the fiber-reinforced composites [13, 14]. Energy dissipation theory has been used for measuring damping capacity based on vibration damping method.

Damping measurement techniques often deal with natural frequency or resonant frequency of a system. The experimental setup to find the vibration is categorized as free vibration (or free decay) and forced vibration. Free-free beam technique and the piezoelectric ultrasonic composite oscillator technique (PUCOT) are forced vibration techniques. Dynamic mechanical analysis (DMA) uses these techniques to find damping characteristics of the material. However, the instrument is relatively expensive and it cannot be operated at higher frequency and low amplitudes where more information from the tested materials can be obtained [15].

Guan and Gibson [16] have developed micromechanical models for damping in woven fabric-reinforced polymer matrix composites. Where as many other researchers has published results for continuous FRP composites that show damping characteristics of composite material come from microplastic or viscoelastic phenomena associated with the matrix and slippage at the interface between the matrix and the reinforcement [17, 18].

The article presents the methodology to find viscous damping, which is the dominant mechanism in FRP composites vibrating at small amplitudes. A hybrid method employing combined finite element and frequency response has been developed to measure the damping properties of composite material.

2. Experimental setup

2.1 Ultrasonic pulse generator

In this work the specimens are carbon fiber/epoxy (CFRP) and glass fiber/epoxy (GFRP) laminates. The laminate contains woven fiber with 12 plays with average epoxy layer thickness of 0.2 mm. **Figure 1** shows the specimen with the dimensions

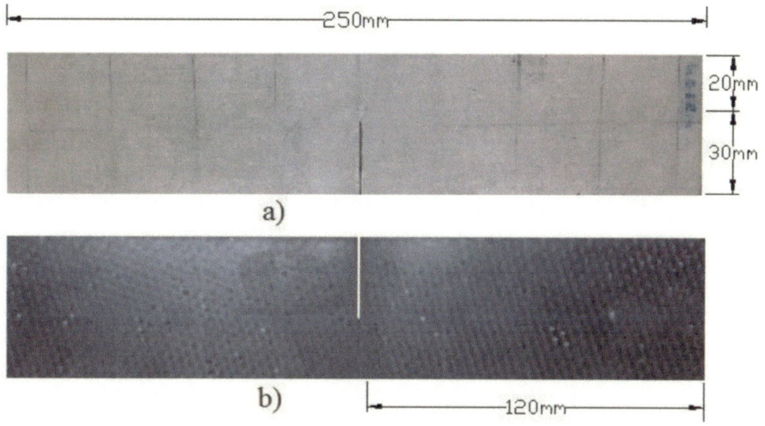

Figure 1.
Damaged specimen: (a) GFRP and (b) CFRP.

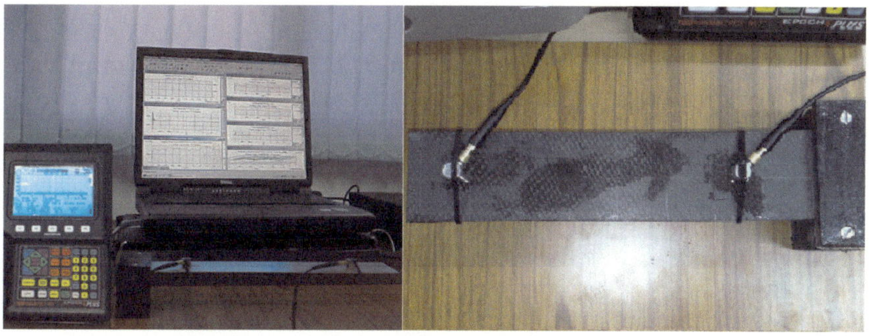

Figure 2.
Experimental setup.

250 × 50 × 2 mm in damaged state where a cut (30 × 2 × 2 mm) at a distance 120 mm away from free end has been introduced.

Figure 2 shows an experimental setup to develop Lamb waves using ultrasonic equipment NDT™ EPOCH 4PLUS. The specimen is supported as cantilever; pitch-catch radio frequency (RF) test method was used. Dual-element transducers (DIC-0408) where one element transmits and the other element receives the burst of acoustic waves generated in the specimen. 1 kHz to 4 MHz frequency range transducers was used and they were placed 80 mm apart from each other. Honey glycerine couplant made by Panametrics has been used to couple the two sensors on to the test specimen. The couplant helps to transmit a normal incident shear wave to propagate across the test piece between the transducer tips. Scanview plus™ software was used to acquire and process the data obtained from the test specimens [19, 20].

2.2 Generation of optimal Lamb wave

Acoustic impedance of the material plays important role in deciding the selection of transducer frequency, low impedance require lower frequency transducers. The materials such as carbon fiber or glass fiber have low impedance thus low frequency transducers will be used to generate Lamb waves. Whereas metal skin layers have high impedance therefore higher frequencies transducers are used for thinner and metallic layers.

The Panametrics-NDT™ EPOCH 4 PLUS is used to generate acoustic waves and the device is equipped with four channels. Optimal Lamb wave propagation parameters were arrived through calibration of the device using editable parameters as shown in **Table 1**.

The parameters have been varied one by one and arrived to a conclusion of optimal driving frequency for different specimens. **Figure 3** shows the optimal driving frequency of different materials calibrated through ultrasonic pulse generator test setup. With the help fitted peak value and percentage amplitude at constant gain 55 (db) Lamb waves generation frequency is identified for different materials. **Figure 4** shows the histogram representation of % amplitude of the waveform with a bin range of 0–820 kHz of pulser [21, 22]. The most effective range of frequencies to generate Lamb waves is 140–420 kHz. The frequency range to generate Lamb waves for GFRP is 170–190 kHz where as it is 260–280 kHz for CFRP and for Aluminum (Al) it is 360–380 kHz.

2.3 Material properties

The Young's modulus for the specimens is calculated from Eq. 1, the calibrated experimental setup is used to generate Acoustic Emission (AE) on to the test specimen. AE velocities travelling in the material are determining from the instrument EPOCH 4 PLUS [22, 23].

Editable parameters							
Pulser	Mode Energy Wave Type Frequency	Receiver	Gain Broad band Low pass High pass By pass	Device	Unit Angle Thickness	Waveform	Rang Rectification Offset

Table 1.
EPOCH 4 PLUS parameters.

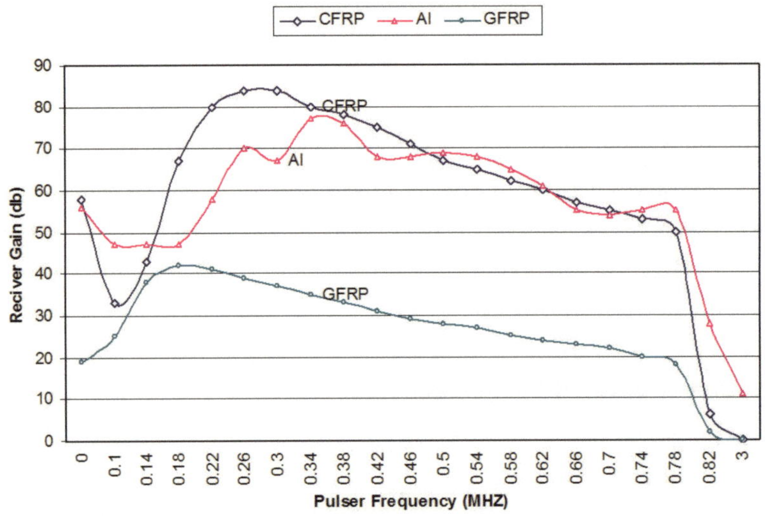

Figure 3.
Optimal driving frequency selection for different materials.

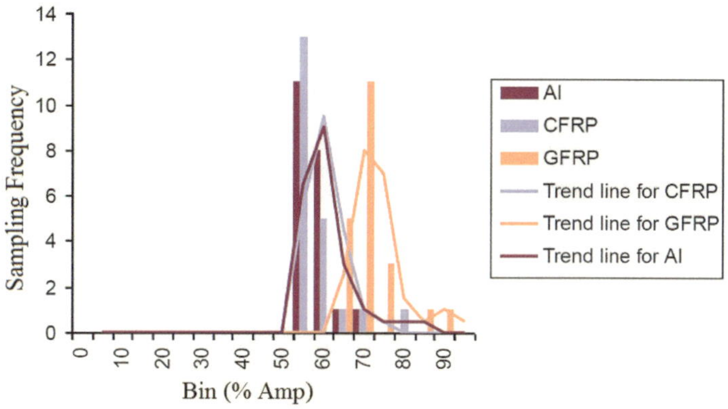

Figure 4.
Histogram representation of % amplitude of waveform at constant gain.

Material	V_T (m/s)	V_L (m/s)	υ_{12}	E_1 (Gpa)	E_2 (Gpa)	υ_{21}	Density (ρ) kg/m³
GFRP	6439	3128	0.346	47.31	11.26	0.081	1853
CFRP	8745	4628	0.307	77.74	21.73	0.086	1400

Table 2.
Material properties arrived from experimental setup.

$$E_1 = \frac{V_T^2\,\rho(1+\nu_{12})(1-2\nu_{12})}{1-\nu_{12}}$$
$$E_2 = \frac{V_L^2\,\rho(1+\nu_{21})(1-2\nu_{21})}{1-\nu_{21}} \quad (1)$$

where V_L is longitudinal velocity of AE and V_T is transverse velocity of AE traveling in the material respectively, ρ is material density and υ_{12} is Poisson's ratio. The material properties of specimens arrived from the experimental setup is presented in **Table 2**.

3. Methodology

In this work a hybrid method has been proposed for identify change in damping capacity of a material using combined finite element and Lamb wave method. The process diagram of the hybrid method for the dynamic mechanical analysis is shown in **Figure 5**.

The group velocity (c_{g_n}), modal frequency (f_n) are determined from Lamb wave model and loss less finite element model respectively.

3.1 Lamb wave model for laminated composite plate

Lamb waves can be of two groups, symmetric and anti-symmetric, these waves propagate independently of the other and boundary conditions of the wave equation are being satisfied by both of them for this problem. Actuating frequency relating the velocity of Lamb wave propagation has been derived in the following section. Dispersion curves of Lamb wave in a particular material, which plot the

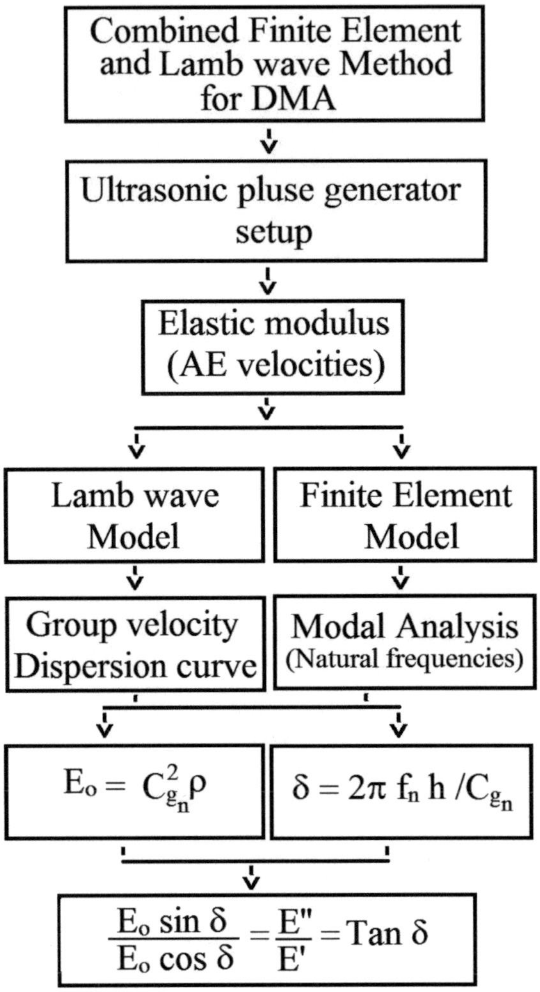

Figure 5.
Methodology chart for dynamic mechanical analysis by hybrid method.

phase and group velocities versus the excitation frequency given by Dalton et al. [24]. The anti-symmetric Lamb wave solution formulated as seen in Eq. 2:

$$\frac{\tan(qh)}{\tan(ph)} = \frac{(k^2 + q^2)^2}{4k^2 qp} \tag{2}$$

where $p^2 = \frac{\omega^2}{c_l^2} - k^2, q^2 = \frac{\omega^2}{c_t^2} - k^2, and\ k = \frac{\omega}{c_{phase}}$
individual laminate stress-strain relationship is given by

$$\begin{bmatrix} \sigma_1 \\ \sigma_2 \\ \tau_6 \end{bmatrix} = \begin{bmatrix} Q_{11} & Q_{12} & 0 \\ Q_{12} & Q_{22} & 0 \\ 0 & 0 & Q_{66} \end{bmatrix} \begin{bmatrix} \varepsilon_1 \\ \varepsilon_2 \\ \gamma_6 \end{bmatrix} \tag{3}$$

where σ is normal stress, τ represent shear stress, ε is normal strain and Υ represent the shear strain. Reduced stiffness components Q_{ij} are defined in terms of the engineering constants as

$$
\begin{aligned}
Q_{11} &= E_1/(1-\upsilon_{12}\upsilon_{21}) \\
Q_{22} &= E_2/(1-\upsilon_{12}\upsilon_{21}) \\
Q_{12} &= \upsilon_{12}E_1/(1-\upsilon_{12}\upsilon_{21})
\end{aligned}
\tag{4}
$$

where E_1 Young's moduli in the longitudinal and E_2 Young's moduli in the transverse directions. The major and minor Poisson's ratios represented by ν_{12} and ν_{21} respectively. The relation between Poisson's ratios in Eq. (4) is given by:

$$
\upsilon_{21} = \frac{E_2}{E_1}\upsilon_{12}
\tag{5}
$$

A_{11} and A_{22}, are in-plane stiffnesses of plate and these are obtained by integrating the Q_{ij} across the thickness of the plate [21]. These stiffness values are given as:

$$
A_{ij} = \int_{-h/2}^{h/2} \left(Q'_{ij}\right)_k dz, i, j = 1, 2,
\tag{6}
$$

where plate thickness is represented by "*h*" and "*k*" *represents* each individual lamina. The transformed stiffness coefficients Q'_{ij} are defined as

$$
\begin{aligned}
Q'_{11} &= m^4Q_{11} + n^4Q_{22} + 2m^2n^2Q_{12} + 4m^2n^2Q_{66} \\
Q'_{22} &= n^4Q_{11} + m^4Q_{22} + 2m^2n^2Q_{12} + 4m^2n^2Q_{66} \\
Q'_{12} &= m^2n^2Q_{11} + m^2n^2Q_{22} + (m^4 + n^4)Q_{12} - 4m^2n^2Q_{66}
\end{aligned}
\tag{7}
$$

where $m = \cos(\theta)$ and $n = \sin(\theta)$, the angle θ is taken positive for counterclockwise rotation and it is considered from the primed (laminate) axes to the unprimed (individual lamina) axes. Q'_{ij} for the 0° and 90° laminas are given by

$$
\begin{aligned}
(Q'_{11})_{0\text{ deg}} &= Q_{11}\ (Q'_{11})_{90\text{ deg}} = Q_{22} \\
(Q'_{22})_{0\text{ deg}} &= Q_{22}\ (Q'_{22})_{90\text{ deg}} = Q_{11} \\
(Q'_{12})_{0\text{ deg}} &= Q_{12}\ (Q'_{12})_{90\text{ deg}} = Q_{12}
\end{aligned}
\tag{8}
$$

The extensional plate mode velocity is related to the in-plane stiffness of a composite [14]. A_{11} and A_{22} are the stiffnesses propagating in the 0° and 90° directions respectively. The relation between extensional plate mode velocity and stiffness is given by:

$$
\text{for } 0°\text{direction } c_t = \sqrt{A_{11}/\rho h}
\tag{9}
$$

$$
\text{for } 90°\text{direction } c_l = \sqrt{A_{22}/\rho h}
\tag{10}
$$

The inplane stiffnesses A_{11} and A_{22} are calculated using Eqs. (4)–(8) by substituting engineering stiffnesses of the composite. The extensional plate mode velocities are substituted into Eq. (2) and it is solved numerically for phase velocity in MathematicaTH. Phase velocity (c_{phase}) is the dependent variable being solved for the independent variable being iteratively supplied is the frequency-thickness product, where ω is the driving frequency in radians. Group velocity dispersion curve, which are derived from the phase velocity curve using Eq. (11):

$$
c_{group} = c_{phase} + \frac{\partial c_{phase}}{\partial k}k = \frac{c_{phase}}{1 - \frac{f}{c_{phase}} \cdot \frac{\partial c_{phase}}{\partial f}}
\tag{11}
$$

where f is the frequency in Hz.

3.2 Finite element model for free vibration of a laminated composite plate

Natural frequency is the phenomenon that occurs with oscillatory motion at certain frequencies known as characteristic values, and it follows well defined deformation pattern known as mode shapes or characteristic modes. The study free vibration is important in finding the dynamic response of elastic structures. It is assumed that the external force vector $\vec{P}$ to be zero and the harmonic displacement as:

$$\vec{Q} = \underline{\vec{Q}} . e^{i\omega t} \tag{12}$$

and the free vibration is given by:

$$\left[[k] - \omega^2 [M]\right] \underline{\vec{Q}} = \vec{O} \tag{13}$$

where $\underline{\vec{Q}}$ is displacement amplitude, $\vec{Q}$ eigen vector and ω denotes the natural frequency of vibration. Eq. (12) is a linear algebraic eigenvalue problem where neither $[k]$ nor $[M]$ is a function of the circular frequency (ω), $\underline{\vec{Q}}$ is nonzero solution therefore the determinant of coefficient matrix $|k| - \omega^2 |M|$ is zero, i.e.,

$$|k| - \omega^2 |M| = 0 \tag{14}$$

where $[k]$ is stiffness matrix and $[M]$ is mass matrix, which are derived through finite element formulation.

Figure 6 shows the plate bending formulation where x, y, and z describes the global coordinate of the plate whereas u, v, and w are the displacements, h represents plate thickness. The xy plane is parallel to the midsurface plane prior to deflection. The displacements in the plate at any point is expressed as

$$u = u(x, y, z) \tag{15}$$

$$v = v(x, y, z) \tag{16}$$

$$w = w(x, y, z) \tag{17}$$

The plane displacement u and v vary through the plate thickness as well as with in the xy-plane while the transverse displacement w remains constant through the plate thickness.

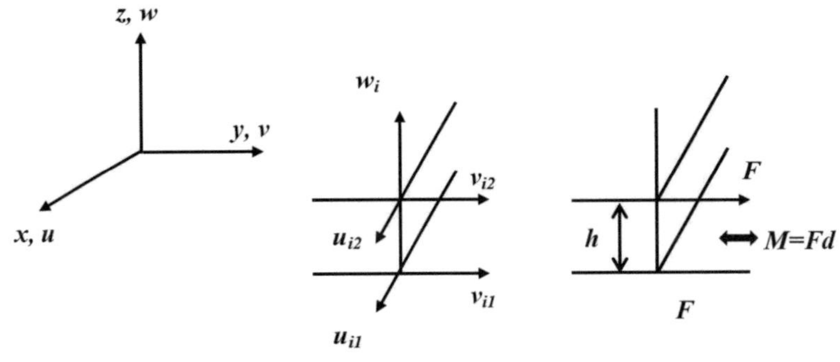

Figure 6.
Plate element with displacement degrees of freedom.

In order to develop shape functions two different interpolations are used one interpolation within the xy-plane and the other in the z-axis. For the xy-plane interpolation, shape function $N_i(x,y)$ are used where subscript i varies depending on the number of nodes on the xy-plane. Shape function $H_j(z)$ is used for interpolation along the z-axis, where subscript j varies depending on the number of nodes along the plate thickness. Since two inplane displacement are functions of x, y, and z, both shape functions are used while the shape functions $N_i(x,y)$ was used for transverse displacement. The mapping of ξ, η-plane onto xy-plane and ζ-axis to z-axis, was done using isoparametric element and the three displacements are expressed as

$$u = \sum_{i=1}^{N_1} \sum_{j=1}^{N_2} N_i(\xi, \eta) H_j(\zeta) u_{ij} \tag{18}$$

$$v = \sum_{i=1}^{N_1} \sum_{j=1}^{N_2} N_i(\xi, \eta) H_j(\zeta) v_{ij} \tag{19}$$

$$w = \sum_{i=1}^{N_1} N_i(\xi, \eta) w_i \tag{20}$$

where N_1 represents the number of nodes in xy-plane (ξ, η-plane) and N_2 represents the number of nodes in z-axis (ζ-axis). The first subscript for u and v denotes the node numbering in terms of xy-plane (ξ, η-plane) and the second subscript indicates the node numbering in terms of z-axis (ζ-axis). Four-node quadrilateral shape function is considered for the xy-plane (ξ, η-plane) interpolation, i.e., $N_1 = 4$ and $N_2 = 2$ that is linear shape function which is considered for the z-axis (ζ-axis) interpolation. Nodal displacement u_{i1} and v_{i1} are displacement on the bottom surface of the plate element and u_{i2} and v_{i2} are displacement on the top surface. As seen in Eqs. (18)–(20), there is no rotational degree of freedom for the present plate bending element were as both bending strain energy and transverse shear strain energy are included.

The relation between bending strains and transverse shear strain with respect to displacements is given by:

$$\{\varepsilon_b\} = \begin{Bmatrix} \varepsilon_x \\ \varepsilon_y \\ \gamma_{xy} \end{Bmatrix} = \begin{bmatrix} \frac{\partial}{\partial x} & 0 & 0 \\ 0 & \frac{\partial}{\partial y} & 0 \\ \frac{\partial}{\partial y} & \frac{\partial}{\partial x} & 0 \end{bmatrix} \begin{Bmatrix} u \\ v \\ w \end{Bmatrix} \tag{21}$$

$$\{\varepsilon_s\} = \begin{Bmatrix} \gamma_{yz} \\ \gamma_{xz} \end{Bmatrix} = \begin{bmatrix} \frac{\partial}{\partial z} & 0 & \frac{\partial}{\partial x} \\ 0 & \frac{\partial}{\partial z} & \frac{\partial}{\partial y} \end{bmatrix} \begin{Bmatrix} u \\ v \\ w \end{Bmatrix} \tag{22}$$

where $\{\varepsilon_b\}$ and $\{\varepsilon_s\}$ are the bending strain and transverse shear strain respectively. The normal strain along the plate thickness ε_z is not considered.

Substitution of Eqs. (18)–(20), into the Eqs. (25) and (26), with $N_1 = 4$ and $N_2 = 2$ gives:

$$\{\varepsilon_s\} = [B_b]\{d^e\} \tag{23}$$

where

$$[B_b] = [[B_{b1}]\ \ [B_{b2}]\ \ [B_{b3}]\ \ [B_{b4}]] \tag{24}$$

$$[B_b] = \begin{bmatrix} H_1\frac{\partial N_i}{\partial x} & 0 & H_2\frac{\partial N_i}{\partial x} & 0 & 0 \\ 0 & H_1\frac{\partial N_i}{\partial y} & 0 & H_2\frac{\partial N_i}{\partial y} & 0 \\ H_1\frac{\partial N_i}{\partial y} & H_1\frac{\partial N_i}{\partial x} & H_2\frac{\partial N_i}{\partial y} & H_2\frac{\partial N_i}{\partial x} & 0 \end{bmatrix} \tag{25}$$

$$\{d^e\} = \left\{ \{d_1^e\}\ \ \{d_2^e\}\ \ \{d_1^e\}\ \ \{d_2^e\} \right\}^T \tag{26}$$

$$\{d_i^e\} = \{ u_{i1}\ \ v_{i1}\ \ u_{i2}\ \ v_{i2}\ \ w_i \} \tag{27}$$

$$\{\varepsilon_s\} = [B_s]\{d^e\} \tag{28}$$

where

$$[B_s] = [\,[B_{s1}]\ \ [B_{s2}]\ \ [B_{s3}]\ \ [B_{s4}]\,] \tag{29}$$

$$[B_{si}] = \begin{bmatrix} N_i\frac{\partial H_1}{\partial z} & 0 & N_i\frac{\partial H_2}{\partial z} & 0 & \frac{\partial N_i}{\partial x} \\ 0 & N_i\frac{\partial H_1}{\partial z} & 0 & N_i\frac{\partial H_2}{\partial z} & \frac{\partial H_2}{\partial y} \end{bmatrix} \tag{30}$$

The constitutive equation is

$$\{\sigma_b\} = [D_b]\{\varepsilon_b\} \tag{31}$$

$$\{\sigma_b\} = \left\{ \sigma_x\ \ \sigma_y\ \ \tau_{xy} \right\}^T \tag{32}$$

$$[D_b] = \frac{E}{1-\upsilon^2}\begin{bmatrix} 1 & \upsilon & 0 \\ \upsilon & 1 & 0 \\ 0 & 0 & \frac{1-\upsilon}{2} \end{bmatrix} \tag{33}$$

For the bending components

$$\{\sigma_s\} = [D_s]\{\varepsilon_s\} \tag{34}$$

where

$$\{\sigma_s\} = \left\{ \tau_{yz}\ \ \tau_{xz} \right\}^T \tag{35}$$

$$[D_s] = \frac{E}{2(1+\upsilon)}\begin{bmatrix} 1 & 0 \\ 0 & 1 \end{bmatrix} \tag{36}$$

where Eq. (33) is for the plane stress condition for the plate bending theory and for a FRP composite, is given by

$$[D_b] = \begin{bmatrix} D_{11} & D_{12} & 0 \\ D_{12} & D_{22} & 0 \\ 0 & 0 & D_{33} \end{bmatrix} \tag{37}$$

In which

$$D_{11} = \frac{E_1}{1 - \upsilon_{12}\upsilon_{21}} \quad (38)$$

$$D_{12} = \frac{E_1 \upsilon_{21}}{1 - \upsilon_{12}\upsilon_{21}} \quad (39)$$

$$D_{22} = \frac{E_2}{1 - \nu_{12}\nu_{21}} \quad (40)$$

$$D_{33} = G_{12} \quad (41)$$

And

$$[D_s] = \begin{bmatrix} G_{13} & 0 \\ 0 & G_{12} \end{bmatrix} \quad (42)$$

Here, the longitudinal direction is represented with 1 and transverse direction is represented with 2 for the FRP composite. Further E_i and G_{ij} are the elastic modulus and shear modulus respectively, whereas υ_{ij} is Poisson's ratio for strain in the j-direction. Five independent material properties will be considered for Eqs. (37)–(42) because of the reciprocal relation

$$\frac{\upsilon_{12}}{E_1} = \frac{\upsilon_{21}}{E_2} \quad (43)$$

The stiffness matrix of the element $\left[k^{(e)}\right]$ is expressed as:

$$\left[k^{(e)}\right] = \int_{\Omega^e} [B_b]^T [D_b][B_b]\,\partial\Omega + \int_{\Omega^e} [B_s]^T [D_s][B_s]\,\partial\Omega \quad (44)$$

where Ω represents the plate domain.
Similarly the mass matrix is given by

$$\left[M^{(e)}\right] = \frac{\rho A t}{9} \begin{bmatrix} 4 & 2 & 1 & 2 \\ 2 & 4 & 2 & 1 \\ 1 & 2 & 4 & 2 \\ 2 & 1 & 2 & 4 \end{bmatrix} \quad (45)$$

where A is area of the element, t is the element thickness and ρ density of material. Natural frequencies are arrived for composite plates from the lossless finite element formulation. The MathematicaTH software has been used to compute the Eigen values using the inputs taken from the experimental data discussed in the previous sections. The analytical model developed was correlated ANSYS model. Block Lancozs method was used to carry out modal analysis in ANSYS. The analysis was done for 30 subsets and shell-190 has been used as meshing element. **Figures** 7 and **8** shows the first and twentieth mode of natural frequency of the undamaged specimen, i.e., GFRP and CFRP respectively and similarly **Figures 9** and **10** shows for damaged specimen.

3.3 Bandwidth method

The damping parameters in FRP composites are based on the energy dissipation mechanism. Vibrational parameters such as frequency and amplitude are used to

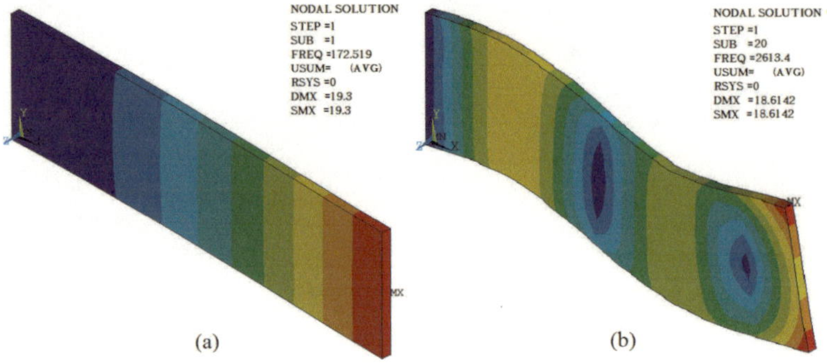

Figure 7.
Undamaged GFRP specimen's first and twentieth mode of natural frequency.

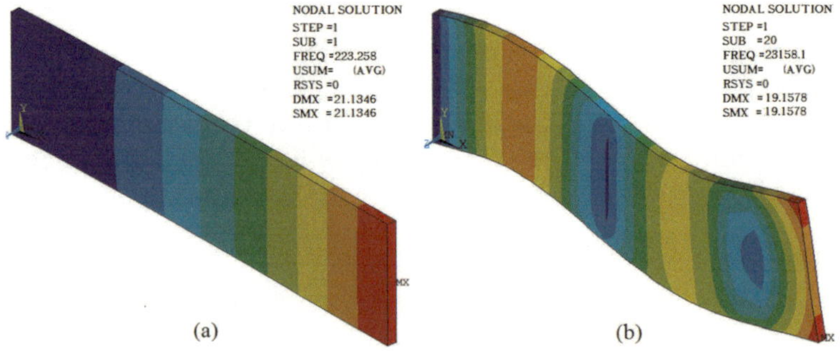

Figure 8.
Undamaged CFRP specimen's first and twentieth mode of natural frequency.

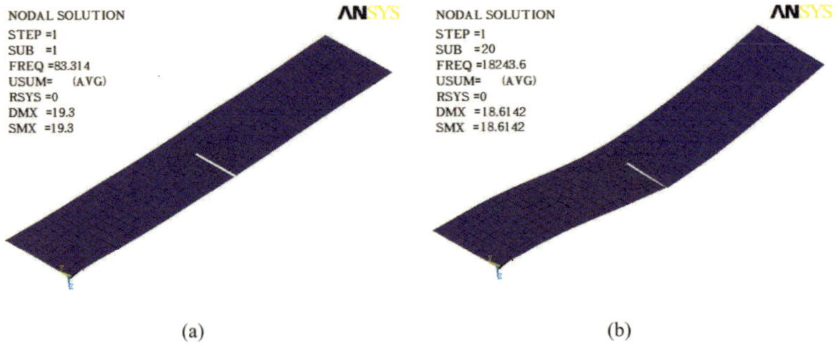

Figure 9.
Damaged GFRP specimen's first and twentieth mode of natural frequency.

determine the dynamic characteristics of a system. The best practice to study vibrational parameters is with nondestructive evaluation. Damping characteristics of a system can be determined by the maximum response, i.e., the response at the resonance frequency as indicated by the maximum value of R_v. **Figure 11** illustrates the Bandwidth method of damping measurement where, damping in a system is indicated by the sharpness or width of the response curve in the vicinity of a resonance frequency ω_r, designating the width as a frequency increment (i.e., $\Delta\omega = \Delta\omega_2 - \omega_1$) measured at the "half-power point" (i.e., at a value $(R/\sqrt{2})$) and the damping ratio ζ can be estimated by using band width in the relation given by

$$\zeta = \frac{\Delta\omega}{2\omega_r} \tag{46}$$

for ith mode damping ratio is given by

$$\zeta_i = \frac{1}{2}\frac{\Delta\omega_i}{\omega_i} \tag{47}$$

The equation of motion of a system with viscous damping, when the excitation is a force $F = F_o \sin \omega t$ applied to the system, is given by

$$mx^{\cdot\cdot} + cx^{\cdot} + kx = F_o \sin \omega t \tag{48}$$

Eq. (48) represents the forced vibration of a damped system and the resulting motion occurs at the forcing frequency ω. The damping coefficient c is greater than zero, leads to change in the phase between the force and resulting motion. The phase change is termed as phase angle δ which is a function of the frequency ratio ω/ω_r and for several values of the fraction of critical damping ζ, given by [25].

$$\boldsymbol{\delta = tan^{-1}\frac{2\zeta(\omega/\omega_r)^3}{1-(\omega^2/\omega_r^2)+(2\zeta\omega/\omega_r)^2}} \tag{49}$$

Figure 10.
Damaged CFRP specimen's first and twentieth mode of natural frequency.

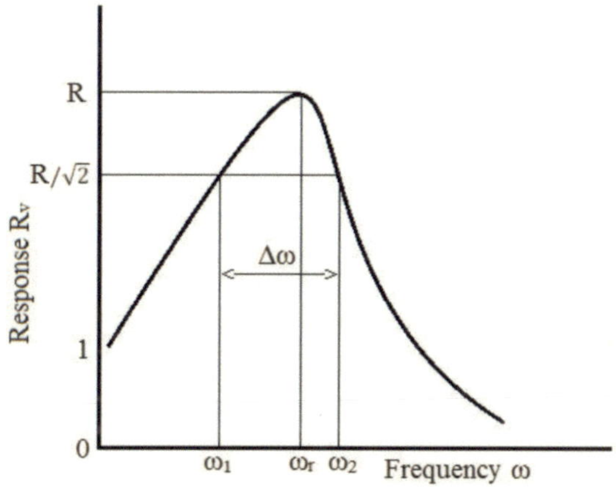

Figure 11.
Response curve showing bandwidth at half-power point.

4. Results and discussion

The phenomenon of change in modal parameters has been used to identify the damage in the specimens. The damage in the specimen is identified by change in damping capacity with respect to undamaged specimen. The first order Lamb wave equation is used to determine the storage modulus. Lamb wave propagating is quite complex to understand, i.e., an increase in modulus slightly speeds the wave velocity. An increase in the density would have the opposite effect slowing wave velocity, as it appears in all the same terms as the modulus but on the reciprocal side of the divisor.

The AE velocities of the specimens were arrived experimentally using ultrasonic pulse generator test setup and the engineering constants, Young's modulus and poison's ratio were calculated. The engineering constants are substituted in Lamb wave model discussed in previous sections for finding dispersion characteristics shown in **Figure 12**. The same material properties are used for finite element model to determine natural frequencies.

The group velocity (c_{g_n}) at natural frequency (f_n) and thickness (h) is substituted in Eq. (50) to determine the phase shift and thus finding material damping capacity ($Tan\ \delta$). Dynamic mechanical analysis can be carried out using the same procedure by getting the E_o value from group velocity dispersion at iteratively supplied frequencies.

$$\delta = 2\pi f_n h / c_{g_n} \tag{50}$$

$$E_o = c_{g_n}^2 . \rho \tag{51}$$

Damping is the term used in vibration and noise analysis to describe any mechanism whereby mechanical energy in the system is dissipated. The damping properties of so-called damping materials, such as elastomeric materials, are usually temperature and frequency dependent, so the experimental determination of damping material properties requires a long and repeating process.

In dynamic mechanical analysis damping measurements is done in temperature sweep mode whereas in this work frequency sweep mode is used. In the present work damping measurements were carried out using combined finite element and Lamb wave method and the results were compared with bandwidth method.

The modal analysis was carried out using developed finite element model and it was correlated with ANSYS. The waveform from the instrument is processed through virtual controlling software and the continuous waveform is subjected to

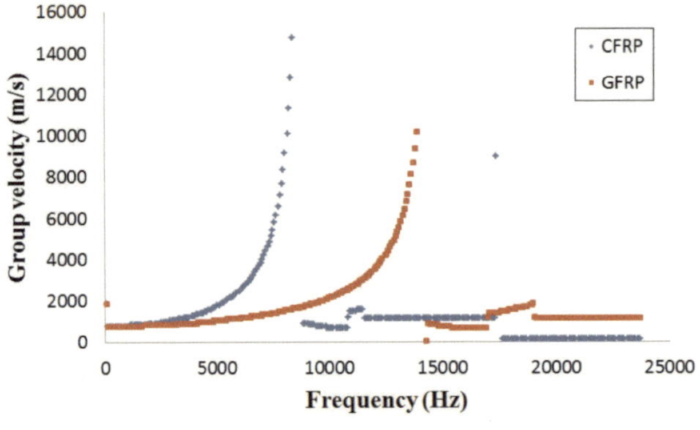

Figure 12.
Lamb wave dispersion curves of CFRP and GFRP.

DOI: http://dx.doi.org/10.5772/intechopen.86134

fast Fourier transform (FFT) which yield a single peak from the calibrated optimal driving frequency, however for a few finite cycles, the FFT appears as a Gaussian curve. The response curve of the undamaged and damaged specimens being tested for damping capacity using bandwidth method is shown in **Figure 13**.

The Lamb wave dispersion curves have been obtained from the iterative supply of the frequency using MathematicaTH code. The group velocity dispersion curve of the specimens used in this research is shown in **Figure 12**. The group velocities and the natural frequencies obtained from modal analysis are used to determine damping capacity at various mode of interest.

Table 3 shows the damping capacities of the undamaged specimen in comparison at critical modes similarly for damaged specimen it has been reported in **Table 4**. It is observed that the natural frequencies of the damaged specimen fell down and the damping capacities have increased slightly with respect to undamaged specimens. **Figure 14** shows the damping capacities and dynamic

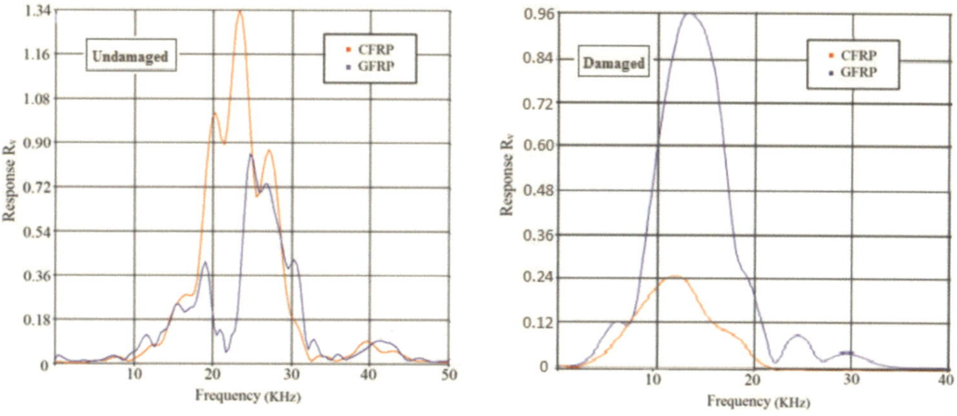

Figure 13.
Response curve of GFRP and CFRP showing bandwidth.

Material	Natural frequency (Hz)	Lamb wave method (Damping capacity)	Mode frequency(Hz)	Bandwidth method (Damping capacity)
GFRP	172(1st)	0.0000493	8692	0.001394
	9970(10th)	0.0012204	11537	0.002427
	21606(20th)	0.0043671	19248	0.004251
CFRP	223(1st)	0.0000646	8706	0.004569
	11023(10th)	0.0016254	20835	0.021765
	23158(20th)	0.0516527	23164	0.049638

Table 3.
Damping capacity of undamaged test specimens.

Material	Natural frequency (Hz)	Lamb wave method (Damping capacity)	Mode frequency(Hz)	Bandwidth method (Damping capacity)
GFRP	83.314(1st)	0.0001542	3692	0.001632
	5216.32(10th)	0.0021472	6924	0.002951
	18243.6(20th)	0.0048795	14248	0.004793
CFRP	173.256(1st)	0.00013258	4733	0.005042
	9872.93(10th)	0.0023742	12841	0.02635
	20146.4(20th)	0.057483	11457	0.05539

Table 4.
Damping capacity of damaged test specimens.

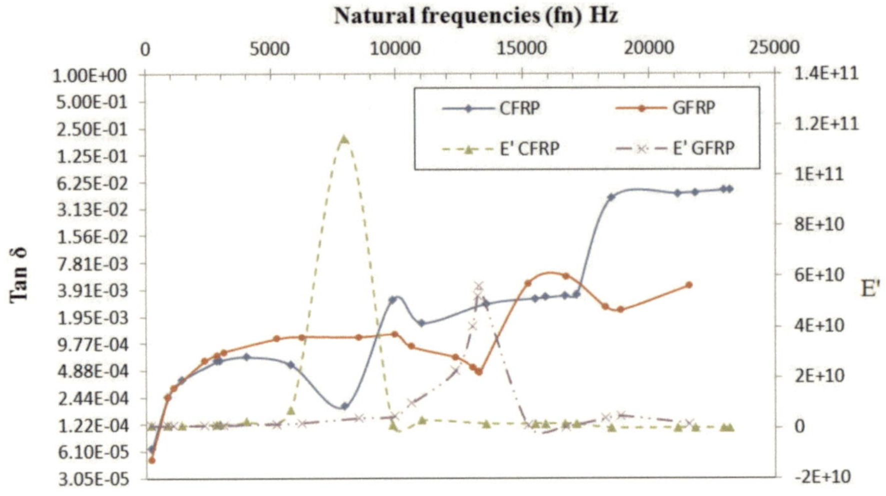

Figure 14.
Damping capacity and dynamic storage modulus for CFRP and GFRP.

storage modulus of the tested specimens with respect to their natural frequencies. The material GFRP and CFRP exhibits similar damping property to a certain range of frequency, and in between 2 and 8 kHz GFRP has better damping property among the two and at higher range of frequencies CFRP is found to be good in damping characteristics.

5. Conclusions

Dynamic mechanical analysis is a technique used to study and characterize damping behavior of materials. It is most useful for studying the viscoelastic behavior of polymers. The tests were conducted on polymer composites CFRP and GFRP laminates in their undamaged and damaged state. A hybrid method has been explored in this work and the materials have been characterized for damping parameters at their mode frequencies. The change in the modal parameters (i.e., natural frequencies and damping capacity) can be used to identify and assesses the health of the structures. It is very advantageous method to obtain damping characteristics of the materials at higher frequency and at relatively low amplitudes.

Author details

Beera Satish Ben[1*] and Beera Avinash Ben[2]

1 National Institute of Technology, Warangal, India

2 Avanthi Institute of Engineering and Technology, Visakhapatnam, India

*Address all correspondence to: satishben@nitw.ac.in

References

[1] Doebling S, Farrar C, Prime M, Shevits D. Damage Identification and Health Monitoring of Structural and Mechanical Systems from Changes in Their Vibration Characteristics: A Literature Review. USA: Los Alamos National Laboratory; 1996. pp. 1-136

[2] Lifshitz J, Rotem A. Determination of reinforcement unbonding of composites by a vibration technique. Journal of Composite Materials. 1969;**3**:412-423

[3] Das S, Saha P, Patro SK. Vibration-based damage detection techniques used for health monitoring of structures. A review. Journal of Civil Structural Health Monitoring. 2016;**6**:477-507

[4] Gillich GR, Praisach ZI. Modal identification and damage detection in beam-like structures using the power spectrum and time-frequency analysis. Signal Processing. 2014;**96**:29-44

[5] Kyriazoglou C, Guild FJ. Finite element prediction of damping of composite GFRP and CFRP laminates—A hybrid formulation-vibration damping experiments and Rayleigh damping. Composites Science and Technology. 2007;**67**:2643-2654

[6] Berthelot JM, Sefrani Y. Damping analysis of unidirectional glass and Kevlar fibre composites. Composites Science and Technology. 2004;**64**: 1261-1278

[7] Berthelot JM. Damping analysis of orthotropic composites with interleaved viscoelastic layers: Modeling. Journal of Composite Materials. 2006;**40**(21): 1889-1909

[8] Berthelot JM, Sefrani Y. Damping analysis of unidirectional glass fiber composites with interleaved viscoelastic layers: Experimental investigation and discussion. Journal of Composite Materials. 2006;**40**(21):1911-1932

[9] Mahi AE, Assarar M, Sefrani Y, Berthelot JM. Damping analysis of orthotropic composite materials and laminates. Composites Part B Engineering. 2008;**39**:1069-1076

[10] Berthelot JM. Damping analysis of laminated beams and plates using the Ritz method. Composite Structures. 2006;**74**:186-201

[11] Berthelot JM, Assarar M, Sefrani Y, Mahi AE. Damping analysis of composite materials and structures. Composite Structures. 2008;**85**:189-204

[12] Chen Y, Gibson RF. Analytical and experimental studies of composite isogrid structures with integral passive damping. Mechanics of Advanced Materials and Structures. 2003;**10**(2): 127-143

[13] Montanari L, Basu B, Spagnoli A, Broderick BM. Damage assessment in a cracked fiber-reinforced cantilever beam using Wavelet-Kurtosis techniques. Key Engineering Materials. 2013;**569**:1226-1233

[14] Sharma DS, Mungla MJ, Barad KH. Vibration-based nondestructive technique to detect crack in multi-span beam. Nondestructive Testing and Evaluation. 2015;**30**:1-21

[15] Botelho EC, Pardini LC, Rezende MC. Damping behavior of continuous fiber/metal composite materials by the free vibration method. Composites Part B Engineering. 2006; **37**:255-264

[16] Guan H, Gibson RF. Micromechanical models for damping in woven fabric-reinforced polymer matrix composites. Journal of

Composite Materials. 2001;**35**(16): 1417-1434

[17] Mu B, Wu HC, Yan A, Warnemuende K, Fu G, Gibson RF, Kim DW. FEA of complex bridge system with FRP composite deck. Journal of Composites for Construction. 2006;**10**(1):79-86

[18] Cao MS, Sha GG, Gao YF, Ostachowicz W. Structural damage identification using damping: A compendium of uses and features. Smart Materials and Structures. 2017; **26**(4):043001

[19] Ben BS, Ratnam C, Ben BA, Yang SH. Ultrasonic based method for damage identification in composites materials. International Journal of Mechanics and Materials in Design. 2012;**8**:297-309

[20] Ben BS, Ratnam C, Ben BA, Yang SH. Ultrasonic based structural damage detection using combined finite element and model Lamb wave propagation parameters in composite materials. International Journal of Advanced Manufacturing Technology. 2013;**67**:1847-1856

[21] Ben BS, Ratnam C, Ben BA, Vikram KA, Yang SH. Damage identification in composite materials using ultrasonic based Lamb wave method. Measurement. 2013;**46**: 904-912

[22] Ben BS, Ben BA, Kweon SH, Yang SH. Structural damping of composite materials using combined FE and Lamb wave method. Structural Engineering and Mechanics. 2014;**51**:1047-1065

[23] Ratnam C, Ben BS, Ben BA. Structural damage detection using combined finite element and model Lamb wave propagation parameters. Journal of Mechanical Engineering Science Part C. 2009;**223**:769-777

[24] Dalton RP, Cawley P, Lowe MJS. The potential of guided waves for monitoring large areas of metallic aircraft fuselage structure. Journal of Nondestructive Evaluation. 2001;**20**: 29-46

[25] Blake RE. Basic vibration theory. In: Piersol AG, Paez TL, editors. Harris' Shock and Vibration Handbook. 6th ed. New York: Mc-Graw Hill; 2010

Chapter 10

Strength Improvement and Stress Analysis of E-Glass Laminated Plates with Circular Notches Using Digital Image Correlation

Abdelhak Khechai, Preetamkumar Marutrao Mohite, Abdelouahab Tati and Mohamed-Ouejdi Belarbi

Abstract

In the current work, the stress concentration and tensile strength degradation of E-glass/epoxy laminates are addressed in the present investigation through a combination of both experimental and numerical studies. The numerical study is performed using finite element method (FEM). The main aim of this work is to improve the ultimate strength of perforated composite plates, by using defense hole system (DHS) technique. The samples are manufactured from commercially available unidirectional (UD) E-glass and clear 1070 resin epoxy. Digital image correlation (DIC) technique is also used to get the full-field surface strain measurements in perforated samples with various open hole diameters and DHS configurations, in order to show their effects on failure strength. Based on the experimental results, the ultimate strength can be improved by introducing two circular auxiliary holes along with the principal stress directions.

Keywords: stress concentration, strength, digital image correlation, notches and defense hole system

1. Introduction

Composite structures have found widespread applications in aerospace and other major industries where weight reduction and directional properties are the main criteria. Circular cutouts are unavoidable in these structures to satisfy the design requirements. However, these cutouts change the mechanical behavior of these structures and produce a high undesirable stress concentration located at the vicinity of these notches. If the material strength is not high enough, failure will undoubtedly occur, usually from the region near the cutout. Therefore, it is mandatory to well identify the stress-strain distributions around the cutout.

Many studies have been done during the past two decades to determine the stress-strain distributions at the circumferential of notches in isotropic and anisotropic structures. A variety of methods have been used to estimate the stress concentration factor (SCF) values, such as exact and approximate analytical analysis. A brief review on current analytical methods for the determination of stress distribution around holes has been given by Sevenois [1].

In structural design field, engineers try to optimize various objectives, such as strength and structural weight, depending upon some requirements. Within the context of optimization, the weight or the strength is the objective function. The structural dimensions such as the thickness, length, or width are the design variables that can be controlled to achieve the best configuration [2]. In the case of composite laminated plates with cutouts, various response mechanisms of these structures are not fully understood and are still topics for continuing research.

Based on Sevenois [1] work, most of the research works to solve the stress concentration problem focus on the stress distributions in orthotropic plates subjected to different loads with different material properties. However, these investigations do not address the problem of whether the stress concentration degree is acceptable for a certain material strength and how one can improve the stress-strain distributions and the strength for these types of structures.

From a design point of view, if there is more than one cutout, the stress concentration at the vicinity of the original notch can be reduced, if one determines the optimum locations of other corresponding holes. This is known as the defense hole theory (DHT). It relies on the following rationale. By introducing small notches (auxiliary holes) on both sides of the main hole, it is possible to smooth the flow of the principal stress paths past the main notch, and this will reduce the SCF around the main notch [3].

This idea is very powerful for reducing the stress concentration. In this context, Erickson and Riley [4] were one of the first investigators to reduce the stress concentration around circular notch in isotropic plates under uniaxial loading. Durelli et al. [5] tried to obtain an ideal boundary of a discontinuity in perforated rectangular plate. They defined this boundary as that boundary along which there is no stress concentration. The ideal design of the boundary of the hole in the rectangular plate reduces the maximum stresses by 26%. On the other hand, the response of orthotropic laminated plates with circular notches has been also studied by Jain [6]. In this study, a FE study was made for reduction of SCF around circular notch in infinite isotropic and orthotropic laminates subjected to uniaxial tension. The SCF was reduced up to 24.4% in isotropic plates and 31% in orthotropic laminates by introducing four auxiliary notches on both sides of the original cutout.

Here, the current investigation addresses the research in the domain of optimization of composites for stress concentration and strength. The main goal of the present experimental and numerical studies is to obtain the best optimal size and position of defense holes for perforated laminates when they are subjected to uniaxial loading condition. For practical industrial applications, the most important characteristic to improve is the strength of the particular structure. Thereby, one of the aims of this study is to contribute to the minimization of the stress concentration and know if there can be a significant improvement in the strength of particular perforated composite laminates. Experimental studies investigated using E-glass/epoxy laminates to validate the improvement of the behavior of perforated laminates with auxiliary holes. Material and specimen preparation steps and different material characterization tests are dealt in detail.

2. Materials and sample preparation

During the present work, samples with different opening diameters were fabricated by the hand lay-up method. A mold release agent is first applied to the mold for getting a high-quality surface finish and facilitates the release of the laminated plates from the metallic mold. When the release agent has cured sufficiently, the UD E-glass fibers are manually placed on the metallic mold. After putting the fibers properly, the resin is applied by brushing. A paint roller is used, in order to distribute the resin uniformly on the metallic mold surface and also to consolidate the lamina, thoroughly wetting the reinforcement and removing the entrapped air. Subsequent layers of the UD glass fibers are added to build the required laminate thickness. The samples were made of four-ply UD E-glass/epoxy lamina. The thickness of each layer is 0.5 mm. The initial materials, E-glass fiber 400 g/m^2 and clear 1070 resin epoxy with a density of 1.15 g/cm^3 and a hardener with a density of 1.02 g/cm^3, were purchased from SF Composites (France).

The completed specimens have been checked to ensure that the final laminates are in good quality without defects and then cut into samples with a length of L = 250 mm and a width of W = 25 mm. These specimens have been cut using a dedicated cutting machine with a diamond-coated blade. Four-layered laminated plates, all in the same direction $[0]_4$, were fabricated in this experimental investigation as perforated specimens.

In order to create circular notches at the center of samples, different sizes of drills were used. On the other hand, to limit the delamination effects at the vicinity of the holes, caused by the drilling process, wooden plates under the samples and a drill machine with a speed of 2300 rad/min were used. Various diameters of drill (2.5, 5, 7.5, and 10 mm) were used in order to obtain various diameter-to-width (D/W) ratios. The main notch is machined by drilling initially a hole of a small diameter and then carefully enlarging it to its final dimension by incremental drill size.

In addition, the following procedure is followed to create different diameters of auxiliary holes in various locations at the vicinity of the main one. The first step is to make transparent papers and fix them onto the laminates using adhesive tape. These papers show the centers of the auxiliary holes. A needle is used in order to mark the center locations of the auxiliary holes on the laminates. Initially, 1 mm diameter drill is used for creating the initial holes. Starting with a small size of a drill improves the accuracy of the locations of the notches. Then, using a drill of sufficient size, a bigger hole centered at the initial hole is created.

Before the testing, the DIC samples were cleaned to remove dirt, and then they were prepared and covered using a black paint and sprayed with a white aerosol to create a random speckle pattern. The samples tested for the present experimental investigation are shown in **Figure 1**.

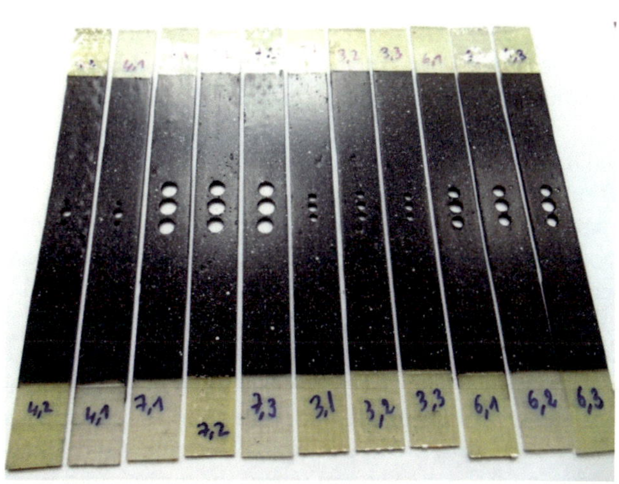

Figure 1.
UD laminated plates with various DHS configurations.

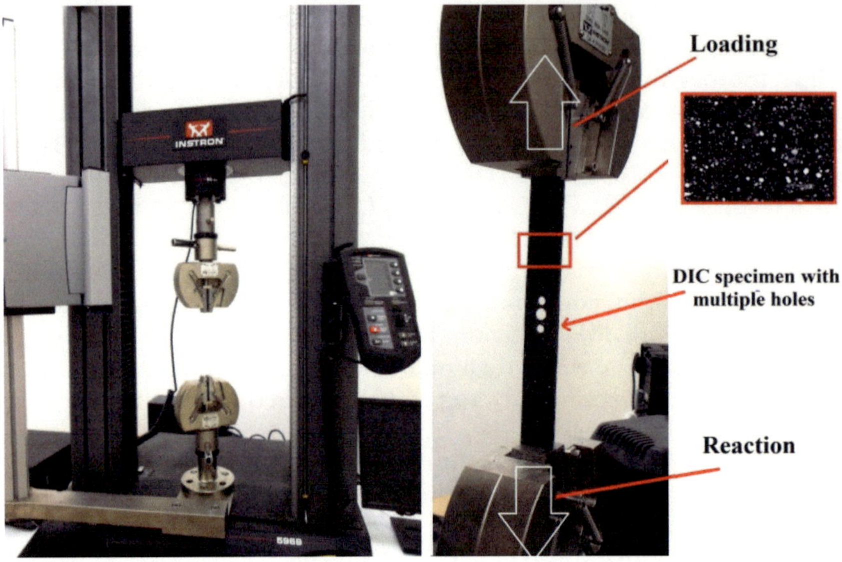

Figure 2.
Experimental setup for the DIC analysis.

In order to obtain the in-plane mechanical properties of the present material, the following ASTM D3039 [7] and ASTM D3518 [8] for tensile and shear properties, respectively, have been used.

The tensile properties of the unnotched samples, such as the laminate Young's modulus E_1 and E_2, Poisson's ratio υ_{12}, and ultimate strength, were measured by static tension testing of longitudinal $[0]_4$ and transverse $[90]_4$ UD samples. The shear modulus of the samples was measured by loading the specimens whose principal axes are on 45°. Four samples were used in the characterization tests.

The INSTRON-5969 testing machine was used in the present study in order to conduct the experimental tests on laminated samples (see **Figure 2**).

The testing machine is connected with a computer in order to record the stress-strain curves during the tensile tests. As shown in **Figure 2**, the samples were illuminated by ordinary white light during the experiments. During the loading process, high-resolution images were taken using a digital camera. The experimental results obtained in the present study were processed with a 2D-DIC MATLAB code [9]. In order to perform the DIC tests, we replicate three experiments for each sample, and the results are averaged. This procedure was repeated for all samples to obtain correct stress distributions and reduce the errors that can be related to the speckle pattern.

Remark: The DHS technique is based on the idea of introducing smaller holes (auxiliary holes) on both sides of the main notch, in order to smooth the flow of the principal stress paths past the main notch, and this will reduce the stress concentration developed around the original notch. This process is similar to the topology optimization technique which is based on logic of "material should be removed from the regions that are less essential for carrying the loads."

Remark: The DIC is one of the powerful noncontact techniques used for measuring the deformations. The DIC technique uses images in order to track the relative displacements of a random speckle pattern point. These displacements are calculated between an undeformed image (reference image) and the current one (the deformed image). In the present work, the authors obtained the full-field strain distributions using a 2D-DIC MATLAB code [9].

DOI: http://dx.doi.org/10.5772/intechopen.87089

3. Finite element modeling

2D finite element models were developed using the open-source FE software FreeFem++ [10]. The models were developed using a linear triangular element (three nodes with 2 degrees of freedom per node), because these elements are more adaptable for meshing plates with circular notches. In order to validate the experimental results, the dimensions and the mechanical properties of the numerical models are chosen to be the same as the experimental specimens (a total length of L = 250 mm and a width of W = 25 mm). In addition, various sizes of hole diameters (2.5, 5, 7.5, and 10 mm) are used in order to obtain different diameter-to-width (D/W) ratios. The length and width of the plates are divided into 70 and 5 elements, respectively.

The FE models of all groups of samples are created, and the stress-strain distributions at the vicinity of notches are obtained. Furthermore, in view of the rapid change in the stress-strain fields around the holes, a higher mesh density with smaller finite elements is adopted and a coarse mesh far from the hole region.

A convergence study is carried out to obtain initial appropriate fine mesh in the open hole zone (the initial mesh size was 30 elements around a hole diameter of 2.5 mm), and then automatic parametric program was developed in order to change the notch size and the mesh refinement automatically, because if one keeps the same element number at the vicinity of the notch boundary, the stress-strain distributions will be affected by changing the notch diameter (see **Figure 3a**). All the numerical models are subjected to a tensile load.

On the other hand, the introduction of the DHS is dependent on the logic of adding auxiliary holes in the areas of low stress near the main cutout. The number of the auxiliary holes in this study is two circular holes (see **Figure 3b**). Various finite element models are also developed for different DHS configurations.

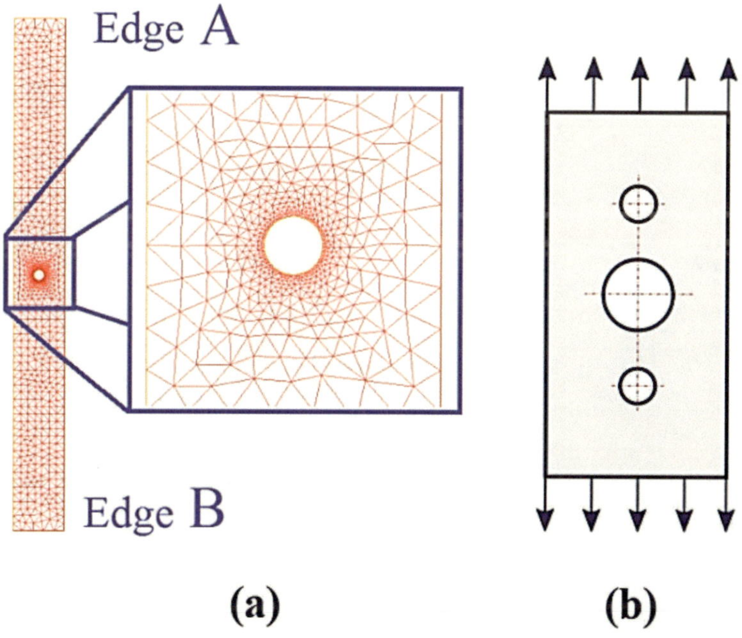

Figure 3.
Finite element models of laminated plates under a tensile loading. (a) A plate with a single hole and (b) a plate with DHS.

4. Results and discussion

In this section, the stress concentration factors and the tensile strengths of laminated plates with different notch diameters are analyzed. The results will be given first for plates with a single hole, and then the same analysis will be done for plates with different DHS configurations.

4.1 Stress concentration and strength of specimens with a single hole (SH)

The performed experimental study aimed to determine the ultimate tensile strength of unnotched/notched samples as well as the stress-strain distributions in laminates weakened by various notch diameters in order to evaluate the stress state. The obtained results from the characterization tests are summarized in **Table 1**.

Table 2 presents the stress concentration and the strength (ultimate stress) obtained experimentally for specimens without notches (unnotched plates) and with a single hole. It can be clearly seen that the SCF and the degradation in strength values are related to the notch dimension. The ultimate strength values vary greatly with the notch size, and the laminate strength decreases as the hole size increases. It turned out that as the notch diameters increased from 2.5 to 10.0 mm, the strength values steadily decreased from 427.57 to 274.86 and the SCF increased from 4.114 to 5.072.

The SCF findings presented in **Table 2** are the average of three values obtained for each sample. One can see that the numerical results are in good agreement compared to the experimental data and the difference between them is considered insignificant.

4.2 Stress concentration and strength of specimens with various DHS configurations

The experimental tensile strength results for the unnotched and notched longitudinal $[0]_4$ laminated specimens with different DHS configurations are shown in **Table 3** and schematically in **Figure 4**. The present DHS configuration is given as $(D/(A, A)/(d, d))$ where D is the main hole diameter, A is the auxiliary hole diameter, and d is the distance between the centers of the main and the auxiliary holes.

Material	Elastic properties			
	E_1 (GPa)	E_2 (GPa)	G_{12} (GPa)	υ_{12}
E-glass/epoxy	22.54	10.94	3.54	0.30

Table 1.
E-glass/epoxy laminate properties.

Material	Ultimate strength	Stress concentration values	
	(MPa)	FEM	Experimental results
Unnotched	527.04	—	—
2.5	427.57	3.756	4.114
5.0	353.54	3.937	4.025
7.5	314.93	4.194	4.799
10.0	274.86	4.601	5.072

Table 2.
Strength and SCF values of unnotched and notched specimens with various notch diameters.

Case		Strength load (MPa)			Improvement
		Unnotched plate	With SH	With DHS	%
Case 1-a	5/(4.5,4.5)/(6,6)	527.04	353.54	413.17	16.87
Case 2-a	5/(4.5,4.5)/(7,7)	527.04	353.54	412.39	16.65
Case 3-a	5/(4,4)/(6,6)	527.04	353.54	386.34	09.27
Case 4-a	5/(2.5,2.5)/(6,6)	527.04	353.54	348.57	−1.41

Table 3.
Experimental strength loads for $[0]_4$ laminates with different DHS configurations.

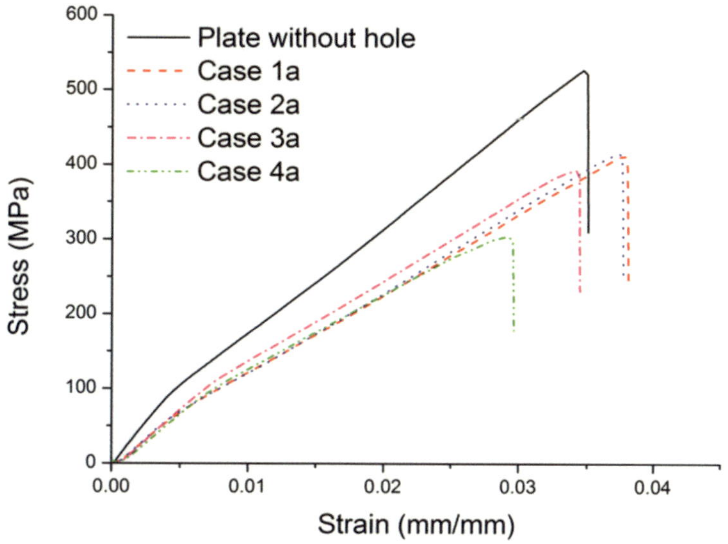

Figure 4.
Typical stress-strain curves for (D = 5 mm) laminated plates with different DHS configurations.

In this experimental investigation, the obtained strength values were compared to plate with a single hole, and the main hole diameter in this case is 5.0 mm.

Figure 4 shows the experimental stress-strain curves, and it can be clearly seen that the strength values were related not only to the main notch diameter but also the auxiliary notch diameters and their locations. The ultimate strength value varies greatly with the auxiliary hole sizes and locations, and the laminate strength increases as the auxiliary notch size increases and is located too near to the main one (see **Table 3**).

4.3 Strain distributions using DIC

As discussed earlier, the DIC technique was used in order to get the strain fields developed in composite samples with different open hole configurations loaded in tension at a rate of 0.5 mm/min. It is shown that the technique provides quantitative information that can be used to identify the strain distribution. A speckle pattern was applied manually on the specimen surface, and the quality of the speckle pattern affects the strain distribution; so it needs to be carefully dropped. The correlation subset size was large enough to ensure that there was a sufficiently distinctive pattern contained in the area used for correlation. In order to calculate the SCF around the

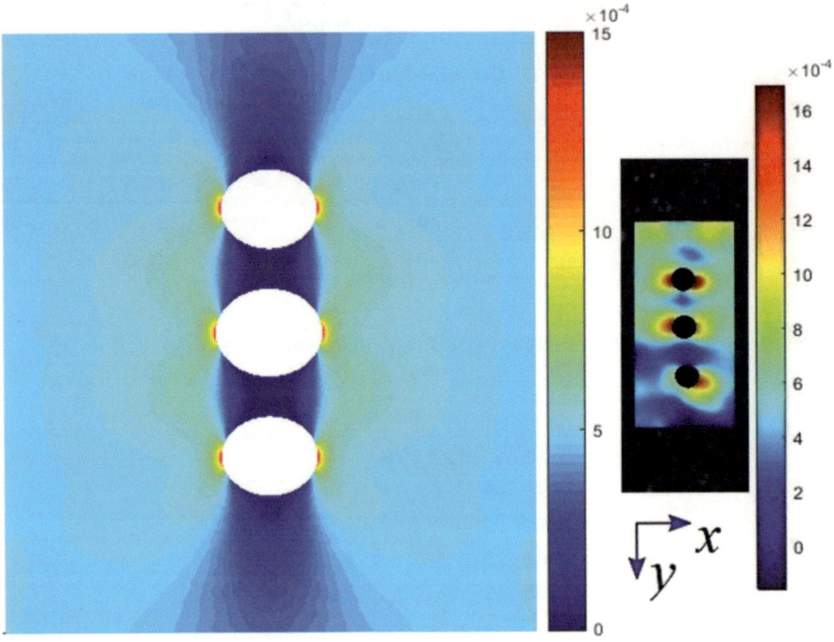

Figure 5.
Finite element and DIC engineering strain fields for $[0]_4$ *notched laminated plate with a 5/(4.5, 4.5)/(7, 7) DHS configuration.*

zone of discontinuities, the stress-strain relationship for planar composite structures was used. DIC technique measurements were carried out in all the samples.

The SCF results tabulated in **Table 2** are actually the average of three values obtained for each specimen. One can see that the experimental results are in good agreement with finite element outputs and the difference between them is insignificant. In addition, one can clearly see that the stress concentration near circular cutouts can be reduced using two auxiliary holes (see **Figure 5**). So, it can be concluded that the stress concentration reduction explains the tensile strength load improvement in specimens with DHS.

5. Conclusion

Notches with different forms and sizes are unavoidable in composite laminates to satisfy the needs of some design requirements. The stress developed around these notches reduces the load-bearing capacity of these composite structures. The accuracy of the analysis of these kinds of structures is based on the choice of an appropriate strategy. In the present investigation, DIC technique was used to the assessment of stress distribution taking place in laminates with different notch sizes loaded in tension. In addition, FEM is also used to validate the distribution of stresses obtained experimentally and investigate the tensile strength degradation in these laminated plates. It has been shown that the ultimate strength degradation and final damage mechanism that appear after the final failure are depending on the open hole size and strain distribution. One can see that the SCF and the damage zones increase and the perforated plate strength decreases with the increase in the hole size. Moreover, to improve the ultimate strength values, the DHS technique was used. DHS is introduced to these structures as a strategy to increase the ultimate strength as well as the weight of the structures. Based on the experimental results, the ultimate strength can be improved by introducing two circular auxiliary holes along the principal stress directions.

Conflict of interest

The authors declare no conflict of interest.

Author details

Abdelhak Khechai[1*], Preetamkumar Marutrao Mohite[2], Abdelouahab Tati[3] and Mohamed-Ouejdi Belarbi[3]

1 Civil Engineering Laboratory, University of Biskra, Algeria

2 Department of Aerospace Engineering, Indian Institute of Technology, Kanpur, India

3 Energy Engineering and Materials Laboratory, University of Biskra, Algeria

*Address all correspondence to: a.khechai@univ-biskra.dz

References

[1] Sevenois R, Koussios S. Analytic methods for stress analysis of two-dimensional flat anisotropic plates with notches: An overview. Applied Mechanics Reviews. 2014;**66**(6):060802. DOI: 10.1115/1.4027562

[2] Gürdal Z, Hafta RT, Hajela P. Design and optimization of laminated composite materials. 1st ed. New York: Wiley & Sons Inc.; 1999

[3] Akour SN, Nayfeh JF, Nicholson DW. Defense hole design for a shear dominant loaded plate. International Journal of Applied Mechanics. 2010;**2**(02):381-398. DOI: 10.1142/S1758825110000548

[4] Erickson PE, Riley WF. Minimizing stress concentrations around circular holes in uniaxially loaded plates. Experimental Mechanics. 1978;**18**(3): 97-100. DOI: 10.1007/BF02325003

[5] Durelli AJ, Brown K, Yee P. Optimization of geometric discontinuities in stress fields. Experimental Mechanics. 1978;**18**(8):303-308. DOI: 10.1007/BF02324161

[6] Jain N. The reduction of stress concentration in a uni-axially loaded infinite width rectangular isotropic/orthotropic plate with central circular hole by coaxial auxiliary holes. International Islamic University Malaysia Engineering Journal. 2012;**12**(6):141-150. DOI: 10.31436/iiumej.v12i6.228

[7] ASTM Committee. Standard test method for tensile properties of polymer matrix composite materials. ASTM D3039/D M; 3039; 2008

[8] ASTM Committee. Standard test method for in-plane shear response of polymer matrix composite materials by tensile test of a ±45 laminate. ASTM D 3518/D; 3518; 2001; (94)

[9] Blaber J, Adair B, Antoniou A. Ncorr: Open-source 2D digital image correlation matlab software. Experimental Mechanics. 2015;**55**(6):1105-1122. DOI: 10.1007/s11340-015-0009-1

[10] Hecht F. New development in Freefem++. Journal of Numerical Mathematics. 2012;**20**(3-4):251-266. DOI: 10.1515/jnum-2012-0013

Chapter 11

Strength Analysis and Variation of Elastic Properties in Plantain Fiber/Polyester Composites for Structural Applications

Christian Emeka Okafor and Christopher Chukwutoo Ihueze

Abstract

Plantain fiber-reinforced composite materials have demonstrated significant properties that are applicable in structural design and development. However, two major concerns arise in relation to the obvious material anisotropy and challenges imposed by structural discontinuity encountered as need for use of fasteners arises. The study assesses the extent of variation of elastic properties (E_x, E_y, G_{xy}, v_{xy}, v_{yx}, m_x, m_y) with fiber orientation using MATLAB functions while considering the extent of variation of the tangential stresses around an idealized functional hole edge. The tensile strength of 410.15 and 288.1 MPa was recorded at 0° fiber orientation angle, while 37.3397 and 33.133 MPa were obtained at fiber orientation angle of 90° for Plantain Empty Fruit Bunch Fiber Composite (PEFBFC) and Plantain Pseudo Stem Fiber Composite (PPSFC), respectively. The tangential stress distribution at hole edge indicated maximum stress value of 119.15 and 100.587 MPa at angular position ø = 90° for PEFBFC and PPSFC, respectively. Judging from various failure indices considered, failure will be initiated at ø = 70° for PEFBFC with stress concentration factor of 2.53 and ø = 65° for PPSFC with stress concentration factor of 2.13, which are less than the stress concentration around the peak stress when angular position is 90°. Both PEFBFC and PPSFC showed similar trends in response to the design scenario considered.

Keywords: elastic properties, structural application, plantain fiber, composites, matrix

1. Introduction

These days several shortcomings have been observed with respect to the utilization of synthetic fiber-reinforced polymers, hence empowering the drive for more utilization of plant fiber composite in structural designs. The primary weaknesses of synthetic fiber-reinforced polymers which includes issue in afterlife disposal and non-biodegradability are completely settled by utilizing plant fibers in polymer reinforcements. Reinforced composites are prime choice for light weight structural designs and automotive body parts assembly.

Extensive literature on recent advancements in reinforced composites and its reliability are reported in classical reports of Dehmous et al. [1], Okafor et al. [2],

Xie and Wang [3], Beaumont et al. [4], Pei et al. [5], Bittrich et al. [6], Prasad et al. [7], Wang et al. [8]. The hygrothermal efficiencies has been reported by Foulc et al. [9], Shettar et al. [10]. In addition, the utilization of reinforced composites predicates the reuse of domestic and agricultural residues. For example, vehicles made with fiber-reinforced composites are lighter and run on smaller engines which produce fewer emissions to the environment. Most items produced using natural fiber composites is a win-win for manufacturers. Most composite material ventures can utilize their genius green item data to build deals because customers comprehend the ecological dangers of synthetic assembling.

Due to scarcely available information regarding some new material response to structural discontinuity, superior properties of those composites are seriously compromised by the utilization of bizarrely enormous factor of safety in design. Accordingly, the quick fate of composite materials as a class of innovative materials may depend more on clear assessment of its performance in various structural design scenario. All inclusive acknowledgment of composites as eco-friendly materials will therefore depend especially on the certainty of the designer and client about the variation of its elastic properties. In a typical fiber-reinforced composites, the polymer matrix serves as a binder and deforming most times for stress distribution purposes. There are different options in the choice of matrix/fibers and the general composition of reinforced composites is shown in **Figure 1**. The figure identified the three major categories of polymers to include biopolymers, thermoplastics and thermosets. Biopolymers are chain like atoms created by organic

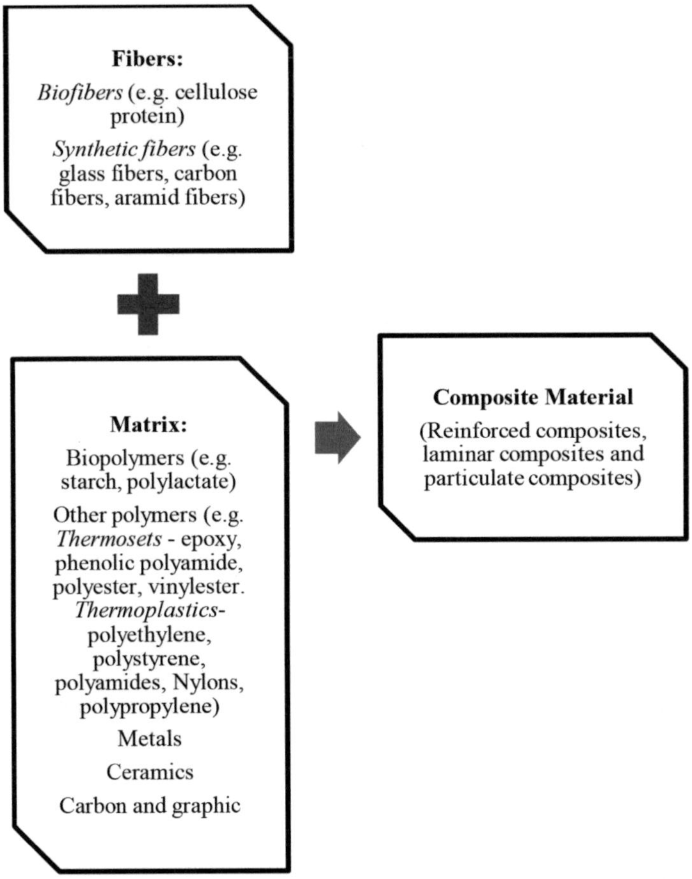

Figure 1.
Composition of reinforced composites.

biomass. Exceptional nontoxicity and biodegradable properties of biopolymers boosts their applications in composites formulation, hardware and restorative gadgets. Fuse of nano-sized support in the biopolymers to improve the properties contributes to the upgrade functional applications of the matrix.

Though thermosets and thermoplastics sound similar, they have very different properties and applications [11]. Thermosets typically changes from fluid to solid state after curing chemical reaction initiated by addition of a catalyst, cross-linker, and curing agent. In the course of the chemical reaction, the material solidifies as a result of cross-linking and formation of longer molecular networks. Subsequently, any further exposure to high heat will cause the material to degrade unlike thermoplastic parts that melts and softens whenever exposed to elevated temperature, thermoset simply become set in their physical and mechanical properties after an initial treatment and therefore are no longer affected by additional heat exposure.

Again, thermoplastics are dissolvable plastics. At temperature above liquefying point, the thermoplastic condenses. The thermoplastic sets once again into solid state when the temperature is reduced and the handling temperature dips under its melting point. This inherent characteristic of thermoplastics enables its softening when heated above its melting point and re-forming as the temperature decreases below the melting point. Most of the times, the expenses of materials for creating thermoset are lower when contrasted with thermoplastic. Also thermoset is regularly simple for wetting the reinforcements and shaping last composites items. Thermoplastics will in general be harder than thermosets and require no refrigeration as uncured thermosets as often as possible do, and can be more effectively be reused and fixed. Elastomers are typically thermosets (requiring vulcanization).

Obviously literature has indicated several approximate relationships between some reinforced composites elastic constants and the homogenized modules of elasticity [12–14]. Also recent research have reported the possibility of measuring variation elastic constants of materials using ultrasonic methods [15–18]. However, scanty research is available on strength analysis and variation of elastic properties in plantain fiber/polyester composites, a gap that the present study seeks to fill.

2. Background to plantain cultivation and utilization as reinforcement in polymer composites

An expanded enthusiasm for the utilization of agricultural wastes in development of reinforced composites has been on the increase. Natural fibers extracted from bio wastes offer a few points of interest over woody biomass, since they are accessible in huge amounts as leftovers and agricultural wastes [19, 20]. The plantain pseudo stem (PPS) and empty fruit bunch (EFB) strands presented in this chapter are agricultural by-products that are biodegradable and locally available from renewable agricultural sources with potentials to contribute to reduction in environmental pollution when utilized in large scale as polymeric reinforcements.

Plantain fruit is one of the staples in Nigeria and it is mainly cultivated in the tropics and ethnic enclaves [21]. It is evaluated that 70 million individuals in West and Central Africa derive most of their nourishment and vitality requirements from plantain fruit plant [22, 23]. Plantain fruit has a fare potential in the light of its huge cultivation and consumption in Nigeria and many other African countries. Akinyemi et al. [24] reported that plantain plant is the third most important plant grown after cassava and yam in Nigeria; collaborating, Kaine and Okoje [25] showed that plantain production is a very profitable enterprise as every ₦1 naira invested in plantain production yields a return on investment of about ₦12.60 kobo. In a study about economics of plantain production Kainga et al. [26] found that the associated high return on investment and short maturity period for plantain contributes to its massive cultivation in Nigeria.

Africa cultivates over 50% of worldwide production of plantain and Nigeria is one of the biggest plantain producing nations in the planet. Therefore the interest in plantain plant fiber for polymer reinforcement was as a result of its abundance and accessibility as it is evaluated that over 15.07 million tons of plantain fruit is produced each year in Nigeria with about 2.4 million metric tons produced from southern Nigeria [27, 28]. Plantain fiber also satisfied over 50% conditions for eco-friendly materials as shown in **Figure 2** and can make strong reinforcement in composites. A composite which can be characterized as a physical blend of at least two unique materials, has properties that are commonly superior to those of any of the establishing materials. It is important to utilize blends of materials to tackle issues in light of the fact that any one material alone cannot suffix effectively in eco-friendly materials technology at an acceptable performance [29].

Cadena Ch et al. [30] and Adeniyi et al. [31] orchestrated the potentials of natural fibers from plantain pseudo stem for use in fiber-reinforced composites. It is therefore important to assess the extent of variation of elastic properties in plantain fiber-reinforced polyester composites to guard against out of plane failure during structural applications. Unfortunately most studies involving plantain fiber-reinforced composites has dwelt on assessment of tensile, flexural and hardness properties [32], optimization of hardness strengths [33], effect of water and organic extractives removal [34], effects of fiber extraction techniques [35], optimization of flexural strength [36], compressive and impact strength evaluation [37–39], effect of high-frequency microwave radiation [40], effect of chemical treatment on the morphology [41], implications of interfacial energetics on mechanical strength [42]. Although Ihueze, Okafor and Okoye [43] has reported the longitudinal (1) and transverse (2) properties of plantain fiber-reinforced composites in **Figure 1**, there is still need to

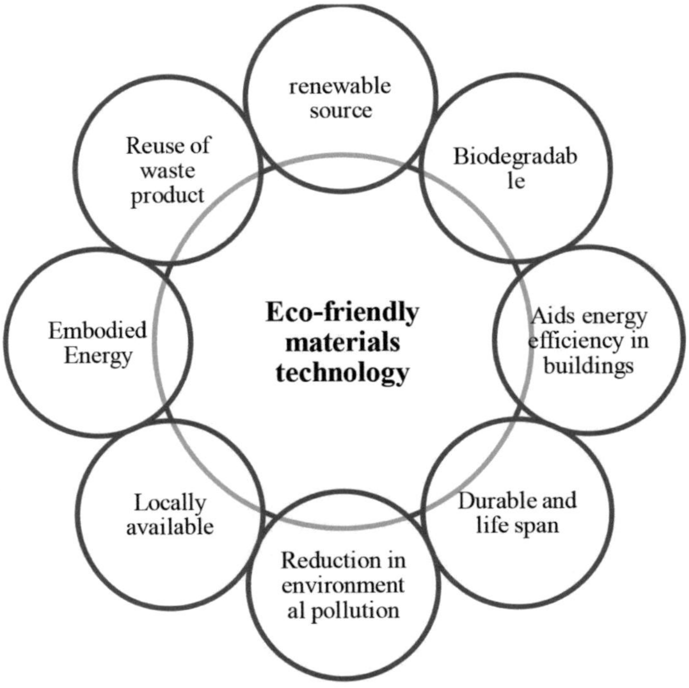

Figure 2.
Properties of eco-friendly building materials.

establish the essential elastic constants at directions other than the material axis directions 1–2.

The present research efforts will further drive the interests of structural designers in the use of plantain fiber-reinforced composites because the superior strength of materials are rarely utilized to full as a result of incomplete knowledge of elastic properties which are related to various fundamental solid-state characteristics of the composites. In essence, the elastic constants of plantain fiber-reinforced composites is expected to describe the material response to external stressor and provide useful information about bonding characteristics and structural stability. Kenedi et al. [44] assessed the orthotropic elastic properties in a sandwiched composites laminates and proposed models for estimating the orthotropic elastic properties of composite materials. Hwang and Liu [45] reported that elastic modulus and Poisson's ratio vary significantly with different braid angles in carbon fabric/polyurethane composites. Ren et al. [46] reported that elastic modulus and tensile strengths are overly dependent on the angles of fiber orientation. Kumar et al. [47] studied the influence of $\pm 0°$, $\pm 10°$, $\pm 30°$, $\pm 40°$, $\pm 45°$, $\pm 55°$, $\pm 65°$, $\pm 75°$, and $\pm 90°$ angle ply on mechanical properties of glass-polyester composite laminate and found that that glass/polyester with 0° fiber orientation angle yields' high strength. Cordin et al. [48] experimentally examined the effect of $0°, \pm 22.5°, \pm 45°, \pm 67.5°$ and 90° fiber orientation angles on the mechanical properties of polypropylene-lyocell composites. Ihueze et al. [49] optimally determined the tensile strengths of plantain fiber-reinforced composites considering 30°, 45° and 90° fiber orientation angles. The application of these previous studies are limited to fiber orientation angles studied, however failure may be initiated from angles other than those considered hence the need to verify the variation of important elastic constants within a wide range of fiber orientation coverage are necessary.

Additionally, researchers have provided various theoretical strategies for determination of elastic constants in reinforced composites using software codes to cover the wide range of fabricating conditions. Jules et al. [50] ascertained the effect of fibers orientation on the predicted elastic properties of long fiber composites using Monte-Carlo simulation to assign the in plane and out of plane orientation values. Venetis and Sideridis [51] developed a model to find the approximate elastic constants in unidirectional fiber-reinforced composite materials in terms of the constituent material properties. Cuartas [52] theoretically determined the elastic properties in CFRP composites which compared favorably with other methods based on tensile tests and ultrasonic characterization. Rahmani et al. [12] found that MATLAB codes are capable of predicting the elastic constants of composites with reasonable confidence.

3. Mathematical framework for assessment of extent of variation of elastic properties in plantain fiber-reinforced polyester composites

One significant property of composite materials is their plainly visible macroscopic anisotropy, which means that the properties estimated in the longitudinal direction are by far not the same as those measured in transverse direction. There are no material planes of symmetry, and normal loads create both normal strains and shear strains. This anisotropic characteristic of reinforced composites results in low mechanical properties in the out-of-plane orientation where the matrix carries the primary load. Consequently the application of reinforced composites is limited in scenarios prone to complex load paths such as lugs and fittings [53].

By implication any endeavor to comprehend the structural application of plantain fiber-reinforced polyester composite must assess the inborn anisotropy.

Composites are a subclass of anisotropic materials that are delegated orthotropic. Orthotropic materials have properties that are unique in three directions with perpendicular axes of symmetry. In this way, orthotropic mechanical properties depend heavily on fiber orientation. An orthotropic ply is thus defined as that having two different material properties in two mutually perpendicular directions at a point and the two mutually perpendicular directions also form the planes of material properties symmetry at the point.

3.1 Determination of reduced stiffness matrix and compliance matrix

Considering two possible loading conditions of longitudinal (direction 1) and transverse (direction 2) in the matrix as shown in **Figure 3**, the resulting direct strains from Hooks law are respectively $e_1 = \frac{-v_{21}\sigma_1}{E_1}$ and $e_2 = \frac{-v_{12}\sigma_1}{E_1}$ where v_{12} = major Poisson's ratio and v_{21} = minor Poisson's ratio.

Hence the application of both direct stresses σ_1 and σ_2 will yield corresponding strains as follows:

$$e_1 = \frac{\sigma_1}{E_1} - \frac{v_{21}\sigma_2}{E_2} \tag{1}$$

$$e_2 = \frac{\sigma_2}{E_2} - \frac{v_{12}\sigma_1}{E_1} \tag{2}$$

Putting Eqs. (1) and (2) in a matrix form, yields

$$\begin{Bmatrix} e_1 \\ e_2 \end{Bmatrix} = \begin{Bmatrix} \frac{1}{E_1} & \frac{-v_{21}}{E_2} \\ \frac{-v_{12}}{E_1} & \frac{1}{E_2} \end{Bmatrix} \begin{Bmatrix} \sigma_1 \\ \sigma_2 \end{Bmatrix} \tag{3}$$

$$\begin{Bmatrix} \sigma_1 \\ \sigma_2 \end{Bmatrix} = \begin{Bmatrix} \frac{E_1}{1 - v_{12}v_{21}} & \frac{v_{21}E_1}{1 - v_{12}v_{21}} \\ \frac{v_{12}E_2}{1 - v_{12}v_{21}} & \frac{E_2}{1 - v_{12}v_{21}} \end{Bmatrix} \begin{Bmatrix} e_1 \\ e_2 \end{Bmatrix} \tag{4}$$

Eq. (3) is symetric about the loading diagonal such that

$$\frac{-v_{21}}{E_2} = \frac{-v_{12}}{E_1} \tag{5}$$

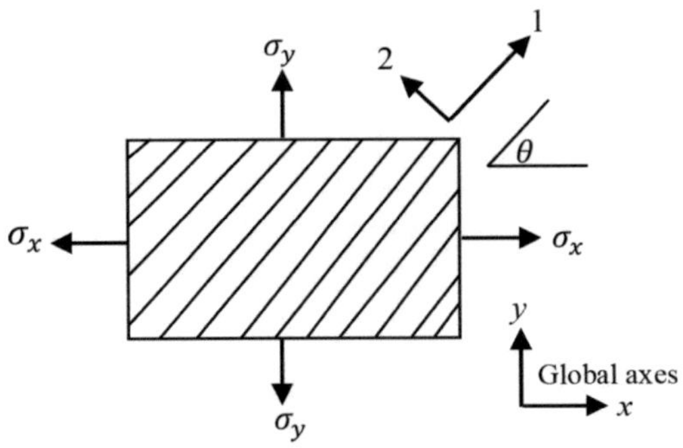

Figure 3.
Stressed single thin composite lamina. From Ref. [43].

A combined effect of shear and direct stresses gives the reduced stiffness matrix as in Eq. (6) and reduced compliance matrix as in Eq. (7)

$$\begin{Bmatrix} \sigma_1 \\ \sigma_2 \\ \tau_{12} \end{Bmatrix} = \begin{Bmatrix} \frac{E_1}{1-v_{12}v_{21}} & \frac{v_{21}E_1}{1-v_{12}v_{21}} & 0 \\ \frac{v_{12}E_2}{1-v_{12}v_{21}} & \frac{E_2}{1-v_{12}v_{21}} & 0 \\ 0 & 0 & G_{12} \end{Bmatrix} \begin{Bmatrix} e_1 \\ e_2 \\ e_{12} \end{Bmatrix} \tag{6}$$

$$\begin{Bmatrix} e_1 \\ e_2 \\ e_{12} \end{Bmatrix} = \begin{Bmatrix} \frac{1}{E_1} & \frac{-v_{21}}{E_2} & 0 \\ \frac{-v_{21}}{E_1} & \frac{1}{E_2} & 0 \\ 0 & 0 & \frac{1}{G_{12}} \end{Bmatrix} \begin{Bmatrix} \sigma_1 \\ \sigma_2 \\ \tau_{12} \end{Bmatrix} \tag{7}$$

Minor Poisson's ratio is the strain resulting from a stress in the axial direction, Ihueze et al. (2013) calculated the major Poisson's ratio v_{12} for plantain fiber/ polyester composites. However, there is need to further assess the minor Poisson's ratio v_{21} using Eq. (5) and **Table 1** as follows

$$v_{21,\text{ PEFBFRP}} = \frac{E_2 v_{12}}{E_1} = \frac{7030.962 * 0.38}{14,922} = 0.179$$

$$1 - v_{12}v_{21} = 1 - 0.38 * 0.179 = 1 - 0.068 = 0.93$$

$$v_{21,\text{ PPSFC}} = \frac{E_2 v_{12}}{E_1} = \frac{6817.175 * 0.29}{13027.5} = 0.152$$

$$1 - v_{12}v_{21} = 1 - 0.29 * 0.152 = 1 - 0.068 = 0.956$$

The reduced stiffness matrix (ℵ) for PEFBFC and PPSFC is obtained from Eq. (6).

For PEFBFC

$$\frac{E_1}{1-v_{12}v_{21}} = \frac{14922}{1-0.38*0.179} = 16011.07$$

$$\frac{v_{21}E_1}{1-v_{12}v_{21}} = 0.179 * 16011.07 = 2865.98$$

$$\frac{E_2}{1-v_{12}v_{21}} = \frac{7030.962}{1-0.38*0.179} = 7544.113$$

Composites	Properties								
	S_{u1} (MPa)	S_{u2} (MPa)	S_y (MPa)	E_1 (MPa)	E_2 (MPa)	E (MPa)	v_{12}	τ_{max} (MPa)	G_{12} (MPa)
PEFBFC	410.15	37.3397	33.69	14,922	7030.962	9990.10	0.38	19.3100	3622.99
PPSFC	288.10	33.1330	29.24	13027.5	6817.175	9146.305	0.29	15.5700	3332.835

S_{u1}, S_{u2} are tensile strengths in the longitudinal and transverse directions respectively.

Table 1.
Evaluated mechanical properties of plantain fibers and plantain fibers reinforced polyester composites. From Ref. [43].

$$\aleph_{PEFBFC} = \begin{Bmatrix} 16011.07 & 2865.98 & 0 \\ 2865.98 & 7544.113 & 0 \\ 0 & 0 & 3622.99 \end{Bmatrix} \text{MPa}$$

For PPSFC

$$\frac{E_1}{1 - v_{12}v_{21}} = \frac{13027.5}{1 - 0.29 * 0.152} = 13628.23$$

$$\frac{v_{21}E_1}{1 - v_{12}v_{21}} = 0.152 * 13628.23 = 2071.49$$

$$\frac{E_2}{1 - v_{12}v_{21}} = \frac{6817.175}{1 - 0.29 * 0.152} = 7131.53$$

$$\aleph_{PPSFC} = \begin{Bmatrix} 13628.23 & 2071.49 & 0 \\ 2071.49 & 7131.53 & 0 \\ 0 & 0 & 3332.835 \end{Bmatrix} \text{MPa}$$

The reduced compliance matrix (β) is obtained from Eq. (7).
For PEFBFC

$$\frac{1}{E_1} = \frac{1}{14922} = 6.7 \times 10^{-5} 1/\text{MPa}$$

$$\frac{-v_{21}}{E_2} = \frac{-0.179}{7030.962} = -2.5 \times 10^{-5} 1/\text{MPa}$$

$$\frac{1}{E_2} = \frac{1}{7030.962} = 1.4 \times 10^{-4} 1/\text{MPa}$$

$$\frac{1}{G_{12}} = \frac{1}{3622.99} = 2.8 \times 10^{-4} 1/\text{MPa}$$

$$\beta_{PEFBFC} = \begin{Bmatrix} 6.7 \times 10^{-5} & -2.5 \times 10^{-5} & 0 \\ -2.5 \times 10^{-5} & 1.4 \times 10^{-4} & 0 \\ 0 & 0 & 2.8 \times 10^{-4} \end{Bmatrix} 1/\text{MPa}$$

For PPSFC

$$\frac{1}{E_1} = \frac{1}{13027.5} = 7.7 \times 10^{-5} 1/\text{MPa}$$

$$\frac{-v_{21}}{E_2} = \frac{-0.152}{6817.175} = -2.2 \times 10^{-5} 1/\text{MPa}$$

$$\frac{1}{E_2} = \frac{1}{6817.175} = 1.5 \times 10^{-4} 1/\text{MPa}$$

$$\frac{1}{G_{12}} = \frac{1}{3332.835} = 3.0 \times 10^{-4} 1/\text{MPa}$$

$$\beta_{PPSFC} = \begin{Bmatrix} 7.7 \times 10^{-5} & -2.2 \times 10^{-5} & 0 \\ -2.2 \times 10^{-5} & 1.5 \times 10^{-4} & 0 \\ 0 & 0 & 3.0 \times 10^{-4} \end{Bmatrix} 1/\text{MPa}$$

Property	Polyester resin
Density (g/cm^3)	1.2–1.5 (1400 kg/m^3)
Young modulus (MPa)	2000–4500
Tensile strength (MPa)	40–90
Compressive strength (MPa)	90–250
Tensile elongation at break (%)	2
Water absorption 24 h at 20°C	0.1–0.3
Flexural modulus (GPa)	11.0
Poisson's ratio	0.37–0.38
	Plantain pseudo stem fibers
Young modulus (MPa)	23,555
UTS (MPa)	536.2
Strain (%)	2.37
Density (kg/m^3)	381.966
	Plantain empty fruit bunch fibers
Young modulus (MPa)	27,344
UTS (MPa)	780.3
Strain (%)	2.68
Density (kg/m^3)	354.151

Table 2.
Mechanical properties of plantain fibers and polyester resin. From Ref. [43].

3.2 Transformation of elastic constants

The know of the stress-strain relationship in the plantain/polyester composite is completely comprehended by knowing the associated independent engineering elastic constants (E_1, E_2, G_{12} and υ_{12}) as previously determined by Ihueze et al. (2013) as in **Tables 1** and **2**. However, there is need to further establish these elastic properties at different directions of fibers other than directions 1 and 2. Datoo [54] derived various expressions for determination of the elastic properties in the reference axes x-y for any fiber orientation as expressed in Eqs. (8)–(14) where $c = cos\theta$, $s = sin\theta$, E_x, E_y, G_{xy}, υ_{xy}, m_x and m_y are the elastic properties at any fiber orientation θ relative to a reference direction x-y.

$$\frac{1}{E_x} = \frac{c^4}{E_1} + \frac{s^4}{E_2} + c^2 s^2 \left(\frac{1}{G_{12}} - \frac{2\upsilon_{12}}{E_1} \right) \tag{8}$$

$$\frac{1}{E_y} = \frac{s^4}{E_1} + \frac{c^4}{E_2} + c^2 s^2 \left(\frac{1}{G_{12}} - \frac{2\upsilon_{12}}{E_1} \right) \tag{9}$$

$$\frac{1}{G_{xy}} = c^2 s^2 \left(\frac{4}{E_1} + \frac{4}{E_2} + \frac{8\upsilon_{12}}{E_1} \right) + \left(c^2 - s^2\right) \frac{1}{G_{12}} \tag{10}$$

$$\upsilon_{xy} = E_x \left[\left(c^4 + s^4\right) \frac{\upsilon_{12}}{E_1} - c^2 s^2 \left(\frac{1}{E_1} + \frac{1}{E_2} - \frac{1}{G_{12}} \right) \right] \tag{11}$$

The minor Poissons ratio with respect to the material reference axes is obtained from Eq. (5) such that

$$\frac{v_{xy}}{E_x} = \frac{v_{yx}}{E_y} \tag{12}$$

$$m_x = E_x\left[c^3 s\left(\frac{1}{G_{12}} - \frac{2v_{12}}{E_1} - \frac{2}{E_1}\right) - cs^3\left(\frac{1}{G_{12}} - \frac{2v_{12}}{E_1} - \frac{2}{E_2}\right)\right] \tag{13}$$

$$m_y = E_y\left[cs^3\left(\frac{1}{G_{12}} - \frac{2v_{12}}{E_1} - \frac{2}{E_1}\right) - c^3 s\left(\frac{1}{G_{12}} - \frac{2v_{12}}{E_1} - \frac{2}{E_2}\right)\right] \tag{14}$$

3.3 Variation of engineering elastic constants with fiber orientation θ

Considering fiber orientation $\theta°$ ranging from 0° to 90° in increments of 5° the variation of E_x, E_y, G_{xy}, v_{xy}, m_x and m_y has been assessed Plantain Empty Fruit Bunch Fiber Composite (PEFBFC) and Plantain Pseudo Stem Fiber Composite (PPSFC) using Eqs. (8)–(14) and presented in **Tables 3** and **4**, respectively.

Figure 4 shows the variation of elastic modulus with fiber orientation, it can be seen that the highest value of 14,922 and 13027.5 MPa in the reference x-direction (E_x) is attained in the fiber orientation angle 0° for PEFBFC and PPSFC respectively. However as fiber orientation angle changes, there is a sharp drop in the value of elastic modulus in the reference x-direction to a respective lowest value of 7030.96 and 6817.18 MPa as the fiber orientation angle increased to 90°. On the contrary, the lowest value of 7030.96 and 6817.18 MPa were recorded for elastic modulus in the

S/N	$\theta°$	E_x	E_y	G_{xy}	v_{xy}	v_{yx}	m_x	m_y	$\frac{E_x}{E_2}$	$\frac{G_{xy}}{E_2}$
1	0	14922.000	7030.962	3622.990	0.380	0.179	0.000	0.000	2.122	0.515
2	5	14769.770	7053.373	3629.270	0.378	0.180	0.116	0.037	2.101	0.516
3	10	14337.670	7121.323	3647.477	0.372	0.185	0.221	0.073	2.039	0.519
4	15	13690.480	7236.950	3675.727	0.362	0.191	0.304	0.111	1.947	0.523
5	20	12911.620	7403.786	3710.985	0.350	0.201	0.362	0.150	1.836	0.528
6	25	12081.300	7626.692	3749.256	0.336	0.212	0.395	0.190	1.718	0.533
7	30	11262.730	7911.712	3785.946	0.320	0.225	0.405	0.231	1.602	0.538
8	35	10497.990	8265.787	3816.391	0.304	0.239	0.398	0.271	1.493	0.543
9	40	9810.462	8696.213	3836.528	0.288	0.255	0.377	0.310	1.395	0.546
10	45	9209.655	9209.655	3843.571	0.271	0.271	0.346	0.346	1.310	0.547
11	50	8696.213	9810.462	3836.528	0.255	0.288	0.310	0.377	1.237	0.546
12	55	8265.787	10497.990	3816.391	0.239	0.304	0.271	0.398	1.176	0.543
13	60	7911.712	11262.730	3785.946	0.225	0.320	0.231	0.405	1.125	0.538
14	65	7626.692	12081.300	3749.256	0.212	0.336	0.190	0.395	1.085	0.533
15	70	7403.786	12911.620	3710.985	0.201	0.350	0.150	0.362	1.053	0.528
16	75	7236.950	13690.480	3675.727	0.191	0.362	0.111	0.304	1.029	0.523
17	80	7121.323	14337.670	3647.477	0.185	0.372	0.073	0.221	1.013	0.519
18	85	7053.373	14769.770	3629.270	0.180	0.378	0.037	0.116	1.003	0.516
19	90	7030.962	14922.000	3622.990	0.179	0.380	0.000	0.000	1.000	0.515

Table 3.
Variation of engineering elastic constants with fiber orientation θ in PEFBFC.

S/N	$\theta°$	E_x	E_y	G_{xy}	v_{xy}	v_{yx}	m_x	m_y	$\frac{E_x}{E_2}$	$\frac{G_{xy}}{E_2}$
1	0	13027.500	6817.175	3332.835	0.290	0.152	0.000	0.000	1.911	0.489
2	5	12897.620	6830.651	3343.613	0.290	0.154	0.114	0.023	1.892	0.490
3	10	12530.190	6872.014	3375.039	0.291	0.159	0.214	0.047	1.838	0.495
4	15	11983.280	6944.024	3424.350	0.291	0.168	0.293	0.073	1.758	0.502
5	20	11330.960	7051.119	3486.843	0.290	0.180	0.344	0.103	1.662	0.511
6	25	10643.310	7199.198	3555.901	0.287	0.194	0.369	0.136	1.561	0.522
7	30	9974.406	7395.333	3623.334	0.282	0.209	0.371	0.173	1.463	0.532
8	35	9359.192	7647.394	3680.229	0.275	0.224	0.356	0.212	1.373	0.540
9	40	8816.012	7963.509	3718.337	0.265	0.239	0.328	0.252	1.293	0.545
10	45	8351.207	8351.207	3731.758	0.253	0.253	0.292	0.292	1.225	0.547
11	50	7963.509	8816.012	3718.337	0.239	0.265	0.252	0.328	1.168	0.545
12	55	7647.394	9359.192	3680.229	0.224	0.275	0.212	0.356	1.122	0.540
13	60	7395.333	9974.406	3623.334	0.209	0.282	0.173	0.371	1.085	0.532
14	65	7199.198	10643.310	3555.901	0.194	0.287	0.136	0.369	1.056	0.522
15	70	7051.119	11330.960	3486.843	0.180	0.290	0.103	0.344	1.034	0.511
16	75	6944.024	11983.280	3424.350	0.168	0.291	0.073	0.293	1.019	0.502
17	80	6872.014	12530.190	3375.039	0.159	0.291	0.047	0.214	1.008	0.495
18	85	6830.651	12897.620	3343.613	0.154	0.290	0.023	0.114	1.002	0.490
19	90	6817.175	13027.500	3332.835	0.152	0.290	0.000	0.000	1.000	0.489

Table 4.
Variation of engineering elastic constants with fiber orientation θ in PPSFC.

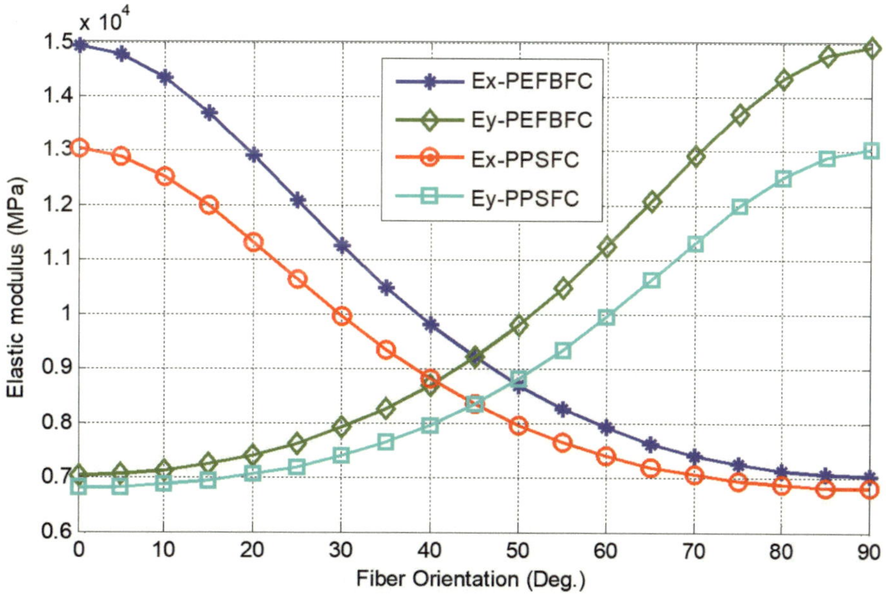

Figure 4.
Variation of elastic modulus with fiber orientation in PEFBFC and PPSFC.

reference *y*-direction (E_y)as the fiber orientation angle increased from 0° to 90° reaching a peak value of 14,922 and 13027.5 MPa for PEFBFC and PPSFC respectively. The implication is that reinforcements are required to be aligned in the direction of applied load [55]. Although Jones [56] intuitively suggested that highest value of material properties may not necessarily occur along the principal material directions, rather it is essential that transverse reinforcement is needed in unidirectional fiber composites which are subjected to multi axial loading [57].

As can be seen in **Figure 5** that shear modulus peaked at 45° fiber orientation and shear modulus was symmetric at about 45° fiber orientation angle for both PEFBFC and PPSFC considered. This implies that the higher in-plane shear resistance is achievable when fiber orientation is 45°. Also the respective minimum value of 3622.99 and 3332.83 MPa at fiber orientation 0° for PEFBFC and PPSFC can be seen to gradually increase to maximum values of 3843.57 and 3731.758 MPa at fiber orientation 45° and then reversed parabolically at 90° where it again reaches to 3622.99 and 3332.84 MPa. Similar trend was obtained by Farooq and Myler [58] who developed efficient procedures for determination of mechanical properties of carbon fiber-reinforced laminated composite panels. This trend in which the value of G_{xy} peaks at 45° fiber orientation angle and lowers at 0° and 90° fiber orientation angle indicates that off-axis reinforcement is very necessary for robust shear stiffness in unidirectional composites [57].

Figure 6 shows variation of Poisson's ratio with fiber orientation, the graph depicts a gradual drop of major Poisson's ratio (v_{xy}) for PEFBFC and PPSFC respectively from 0.38 and 0.29 when fibers are aligned at 0° orientation angle to a lowest value of 0.18 and 0.15 value when fibers were aligned at 90° orientation angle. Additionally, the minor Poissons ratio (v_{yx}) for PEFBFC and PPSFC increased respectively from 0.18 and 0.15 when fibers are aligned at 0° orientation angle to a highest value of 0.38 and 0.29 value when fibers were aligned at 90° orientation angle.

Figure 7 depicts the variation of shear coupling coefficient with fiber orientation, equal magnitude of shear coupling effect was obtained at 45° fiber orientation angle for both PEFBFC and PPSFC considered. Gibson [57] reported that shear coupling coefficient is a measure of the amount of shear strain developed in the *xy* plane per unit normal strain along the direction of the applied normal stress σ_x. **Figure 7** clearly indicate that the maximum value of the shear coupling coefficient in the reference *x*-direction for PEFBFC and PPSFC was attained at 30° fiber orientation angle while the coefficient in the reference *y*-direction for PEFBFC and

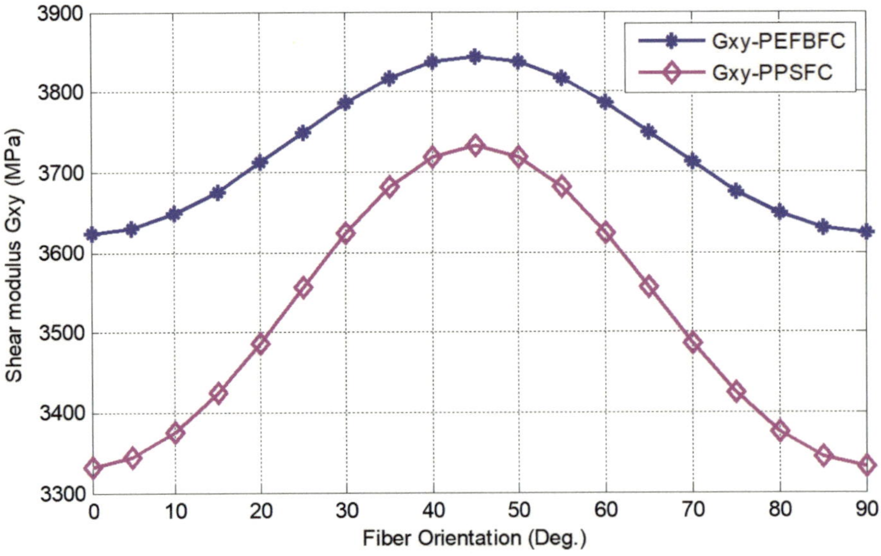

Figure 5.
Variation of shear modulus with fiber orientation in PEFBFC and PPSFC.

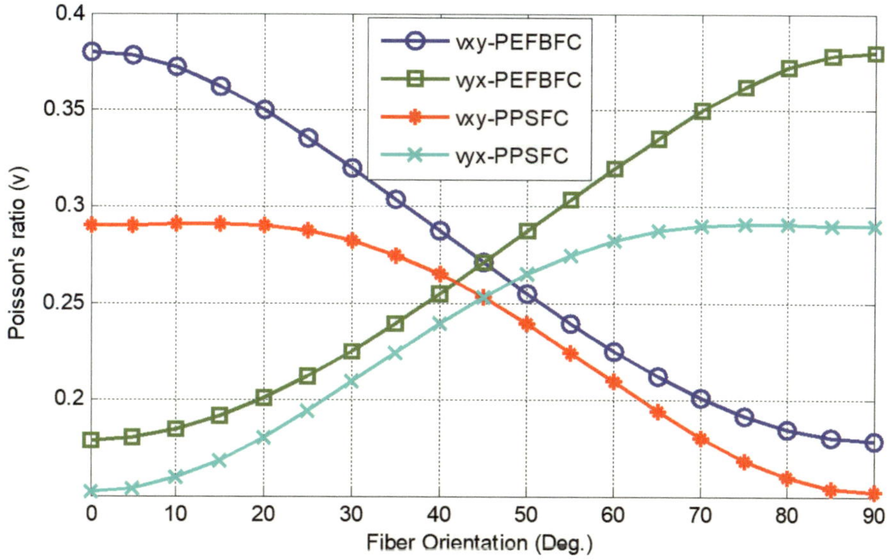

Figure 6.
Variation of Poisson's ratio with fiber orientation in PEFBFC and PPSFC.

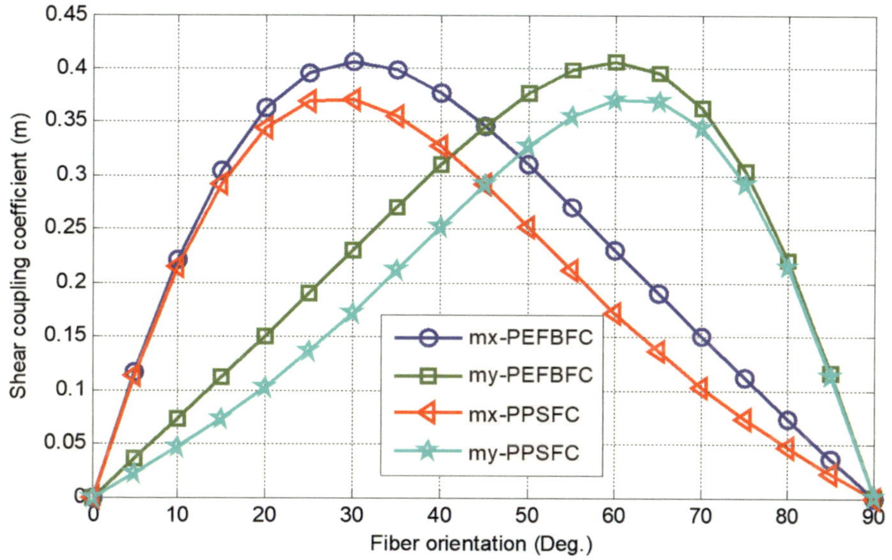

Figure 7.
Variation of shear coupling coefficient with fiber orientation in PEFBFC and PPSFC.

PPSFC was attained at 60° fiber orientation angle. This is an indication that as the shear-coupling ratio increases, the amount of shear coupling increases.

4. Tsai-Hill failure criteria assessment of longitudinal tensile strength

Failure theory is essential in determining whether the composite has failed. Literature review has shown that results of failure prediction depend on failure criterion applied and one major failure criteria used in the industries is Tsai-Hill and failure criteria. Additionally, since composites ultimate tensile strength and strain depend on the fiber orientation, a failure criterion must be used in which the applied stress system is also in material axis [54]. Tsai-Hill theory considers an interaction of the stresses in the fiber direction. It postulates that failure can only occur in reinforced composites when the failure index exceeds 1, hence Eq. (19) must be satisfied to avoid failure.

By considering an arbitrary positive angle θ with reference to the *x*-axis in **Figure 3**, Ihueze et al. [43] transformed the stresses within the global axes (*x*-*y*) into material axes 1–2 as given in Eq. (15)

$$\begin{Bmatrix} \sigma_1 \\ \sigma_2 \\ \tau_{12} \end{Bmatrix} = \begin{bmatrix} c^2 & s^2 & 2sc \\ s^2 & c^2 & -2sc \\ -sc & sc & (c^2 - s^2) \end{bmatrix} \begin{Bmatrix} \sigma_x \\ \sigma_y \\ \tau_{xy} \end{Bmatrix} \tag{15}$$

where $c = cos\theta$ and $s = sin\theta$. Taking longitudinal direction stresses as $\sigma_y = \tau_{xy} = 0$ and thus

$$\sigma_1 = \sigma_x \cos^2\theta \tag{16}$$

$$\sigma_2 = \sigma_x \sin^2\theta \tag{17}$$

$$\tau_{12} = -\sigma_x cos\theta sin\theta \tag{18}$$

Considering Tsai-Hill failure criterion and setting the failure index as 1 for the composite failure to occur:

$$\left(\frac{\sigma_1}{S_{u1}}\right)^2 + \left(\frac{\sigma_2}{S_{u2}}\right)^2 + \left(\frac{\tau_{12}}{\tau_{max}}\right)^2 - \left(\frac{\sigma_1}{S_{u1}}\right)\left(\frac{\sigma_2}{S_{u1}}\right) = 1 \tag{19}$$

Substituting the appropriate value in Eq. (19) we have for PEFBFC

$$\left(\frac{\sigma_x \cos^2\theta}{410.15}\right)^2 + \left(\frac{\sigma_x \sin^2\theta}{37.3397}\right)^2 + \left(\frac{-\sigma_x cos\theta sin\theta}{19.3100}\right)^2 - \left(\frac{\sigma_x \cos^2\theta}{410.15}\right)\left(\frac{\sigma_x \sin^2\theta}{410.15}\right) = 1$$

$$\sigma_{x,\ \mathrm{PEFBFC}} = \sqrt{\frac{1}{\left(cos^4\theta/410.15^2 + \sin^4\theta/37.3397^2 + \cos^2\theta \sin^2\theta/19.3100^2 - \cos^2\theta \sin^2\theta/410.15^2\right)}}$$

And for PPSFC

$$\left(\frac{\sigma_x \cos^2\theta}{288.10}\right)^2 + \left(\frac{\sigma_x \sin^2\theta}{33.1330}\right)^2 + \left(\frac{-\sigma_x cos\theta sin\theta}{15.5700}\right)^2 - \left(\frac{\sigma_x \cos^2\theta}{288.10}\right)\left(\frac{\sigma_x \sin^2\theta}{288.10}\right) = 1$$

$$\sigma_{x,\ \mathrm{PPSFC}} = \sqrt{\frac{1}{\left(\cos^4\theta/288.10^2 + \sin^4\theta/33.1330^2 + \cos^2\theta \sin^2\theta/15.5700^2 - \cos^2\theta \sin^2\theta/288.10^2\right)}}$$

Hence the value for σ_x is then calculated for orientation ranging from 0° to 90° as shown in **Table 5**.

The variation of longitudinal tensile strength with fiber orientation for PEFBFC and PPSFC has been presented in **Figure 8**, it can be seen that the tensile strength equals 410.15 and 288.1 MPa which are the longitudinal tensile strength for PEFBFC and PPSFC respectively when fiber orientation angle is 0°; on the other hand, the tensile strength equals 37.3397 and 33.133 MPa which are the transverse tensile strength for PEFBFC and PPSFC respectively when fiber orientation angle is 90°.

S/N	Orientation θ	PEFBFC	PPSFC
1	0	410.15	288.1
2	5	195.86	152.635
3	10	108.786	86.8937
4	15	75.4556	60.6747
5	20	58.6356	47.3142
6	25	48.8262	39.5106
7	30	42.6439	34.6088
8	35	38.6038	31.4352
9	40	35.9607	29.4005
10	45	34.3043	28.1833
11	50	33.3926	27.5977
12	55	33.0716	27.5303
13	60	33.2311	27.9038
14	65	33.774	28.6492
15	70	34.5913	29.679
16	75	35.5418	30.8595
17	80	36.4447	31.9912
18	85	37.0995	32.8242
19	90	37.3397	33.133

Table 5.
Longitudinal tensile strength variation with fiber orientation angle.

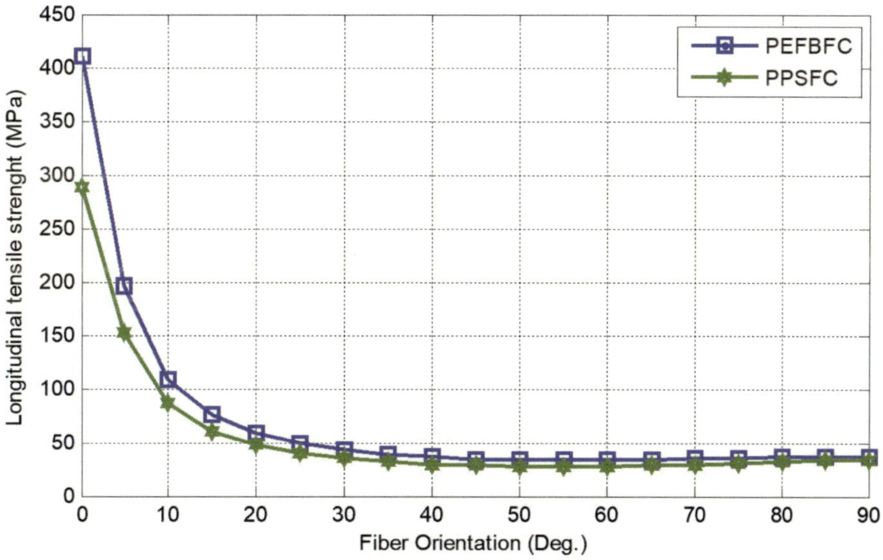

Figure 8.
Variation of longitudinal tensile strength with fiber orientation for PEFBFC and PPSFC.

5. Variation of tangential stress and modulus around a structural discontinuity

Structural discontinuity arising from holes in reinforced composites created for joining or access purposes causes stress concentration at the point of discontinuity [59].

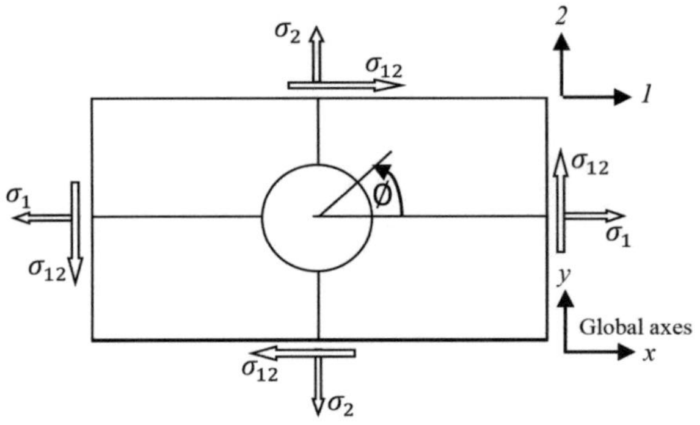

Figure 9.
Depiction of hole in the plantain fiber-reinforced composites sample.

Adequate comprehension of stress redistribution pattern and concentrations is helpful for proficient and safe structural designs. Unlike in ductile materials where stress concentration is of no much ado, plantain fiber-reinforced composites may be sufficiently brittle, hence every form of stress concentration and structural discontinuity has to be properly designed. In a typical scenario where a circular hole is created in the composite as shown in **Figure 9**, assuming no interlaminar stresses exist around the free edge of the hole, the ply is nominally stressed by σ_1, σ_2, σ_{12} some distance away from the hole as indicated. Lekhnitskii [60] derived various useful expressions for stress distribution around holes in a composite plate, the tangential elastic modulus $E_\varnothing$ at an angular position $\varnothing$ is determined using Eq. (20).

$$E_\varnothing = \frac{1}{\left(\frac{sin^4\varnothing}{E_1} + \left[\frac{1}{G_{12}} - \frac{2v_{12}}{E_1}\right] sin^2\varnothing\, cos^2\varnothing + \frac{cos^4\varnothing}{E_2}\right)} \tag{20}$$

Hence the tangential stress $\sigma_\varnothing$ at the perifery of the hole with an angle $\varnothing$ is found from Eq. (21).

$$\sigma_\varnothing = \frac{E_\varnothing}{E_1}(A\sigma_1 + B\sigma_2 + C\sigma_{12}) \tag{21}$$

where

$$A = cos^2\varnothing + (1+p)\, sin^2\varnothing$$

$$B = q\{(q+p)\ cos^2\varnothing - sin^2\varnothing\}$$

$$C = (1+q+p)p\ sin\, 2\varnothing$$

$$p = \sqrt{2(q - v_{12}) + \frac{E_1}{G_{12}}}$$

$$q = \sqrt{\frac{E_1}{E_2}}$$

Using the stress transformation matrix and replacing the axes system 1–2 by radial (r)-tangential ($\varnothing$), we can resolve the tangential stress $\sigma_\varnothing$ back into the material axesin Eq. (22) for proper strength evaluation

$$\begin{Bmatrix} \sigma_r \\ \sigma_\varnothing \\ \tau_{r\varnothing} \end{Bmatrix} = \begin{bmatrix} c^2 & s^2 & -2sc \\ s^2 & c^2 & 2sc \\ sc & -sc & (c^2 - s^2) \end{bmatrix} \begin{Bmatrix} \sigma_x \\ \sigma_y \\ \tau_{xy} \end{Bmatrix} \tag{22}$$

At the edge of the hole, only the tangential stress $\sigma_{\varnothing} > 0$, thus $\sigma_r = \sigma_{r\varnothing} = 0$ in Eq. (22), therefore

$$\sigma_1 = \sigma_x = \sigma_{\varnothing} \sin^2 \varnothing \quad (23)$$

$$\sigma_2 = \sigma_y = \sigma_{\varnothing} \cos^2 \varnothing \quad (24)$$

$$\sigma_{12} = \sigma_{xy} = -\sigma_{\varnothing} \cos \varnothing \sin \varnothing \quad (25)$$

Using the maximum stress criterion, the material will fail when any stress value in the material axes exceeds their respective ultimate strenght. Such that

$$\left| \frac{\sigma_1}{S_{u1}} \right| < 1 \quad (26)$$

$$\left| \frac{\sigma_2}{S_{u2}} \right| < 1 \quad (27)$$

$$\left| \frac{\sigma_{12}}{\tau_{max}} \right| < 1 \quad (28)$$

S/N	$E_\varnothing$	$\sigma_\varnothing$	σ_1	σ_2	σ_{12}	F.I.$_1$	F.I.$_2$	F.I.$_{12}$	Stress conc.
0	7030.962	−23.338	0.000	−23.338	0.000	0.000	0.625	0.000	−0.69
5	7053.373	−22.807	−0.173	−22.634	1.980	0.000	0.606	0.103	−0.67
10	7121.323	−21.211	−0.640	−20.571	3.627	0.002	0.551	0.188	−0.62
15	7236.950	−18.542	−1.242	−17.300	4.636	0.003	0.463	0.240	−0.55
20	7403.786	−14.786	−1.730	−13.056	4.752	0.004	0.350	0.246	−0.43
25	7626.692	−9.917	−1.771	−8.146	3.799	0.004	0.218	0.197	−0.29
30	7911.712	−3.903	−0.976	−2.927	1.690	0.002	0.078	0.088	−0.11
35	8265.787	3.303	1.087	2.217	−1.552	0.003	0.059	0.080	0.10
40	8696.213	11.751	4.855	6.896	−5.786	0.012	0.185	0.300	0.35
45	9209.655	21.484	10.742	10.742	−10.742	0.026	0.288	0.556	0.63
50	9810.462	32.515	19.080	13.434	−16.010	0.047	0.360	0.829	0.96
55	10497.994	44.784	30.050	14.733	−21.042	0.073	0.395	1.090	1.32
60	11262.726	58.103	43.577	14.526	−25.159	0.106	0.389	1.303	1.71
65	12081.304	72.076	59.203	12.873	−27.607	0.144	0.345	1.430	2.12
70	12911.622	86.025	75.962	10.063	−27.648	0.185	0.269	1.432	2.53
75	13690.476	98.951	92.323	6.628	−24.738	0.225	0.178	1.281	2.91
80	14337.671	109.599	106.294	3.305	−18.742	0.259	0.089	0.971	3.22
85	14769.765	116.668	115.782	0.886	−10.130	0.282	0.024	0.525	3.43
90	14922.000	119.152	119.152	0.000	0.000	0.291	0.000	0.000	3.50
95	14769.765	116.668	115.782	0.886	10.130	0.282	0.024	0.525	3.43
100	14337.671	109.599	106.294	3.305	18.742	0.259	0.089	0.971	3.22
105	13690.476	98.951	92.323	6.628	24.738	0.225	0.178	1.281	2.91

Table 6.
Variation of tangential stress, material axis stress and tangential modulus at the edge of material discontinuity in PEFBFC.

The left hand side of Eqs. (26)–(28) represents the failure indices (FI). The maximum failure index (FI) for the applied stress is factored in to obtain the load factor. Due to the inherent material orthotropy, the failure zone of the plantain

S/N	$E_ø$	$\sigma_ø$	σ_1	σ_2	σ_{12}	$F.I._1$	$F.I._2$	$F.I._{12}$	Stress conc.
0	6817.175	−20.978	0.000	−20.978	0.000	0.000	0.633	0.000	−0.72
5	6830.651	−20.459	−0.155	−20.304	1.776	0.001	0.613	0.114	−0.71
10	6872.014	−18.909	−0.570	−18.339	3.234	0.002	0.554	0.208	−0.65
15	6944.024	−16.346	−1.095	−15.251	4.086	0.004	0.460	0.262	−0.56
20	7051.119	−12.791	−1.496	−11.295	4.111	0.005	0.341	0.264	−0.44
25	7199.198	−8.269	−1.477	−6.792	3.167	0.005	0.205	0.203	−0.29
30	7395.333	−2.793	−0.698	−2.095	1.209	0.002	0.063	0.078	−0.10
35	7647.394	3.635	1.196	2.439	−1.708	0.004	0.074	0.110	0.13
40	7963.509	11.024	4.555	6.469	−5.428	0.016	0.195	0.349	0.38
45	8351.207	19.391	9.695	9.695	−9.695	0.034	0.293	0.623	0.67
50	8816.012	28.736	16.863	11.873	−14.150	0.059	0.358	0.909	0.99
55	9359.192	39.014	26.179	12.835	−18.331	0.091	0.387	1.177	1.35
60	9974.406	50.087	37.565	12.522	−21.688	0.130	0.378	1.393	1.73
65	10643.308	61.651	50.640	11.011	−23.614	0.176	0.332	1.517	2.13
70	11330.958	73.175	64.615	8.560	−23.518	0.224	0.258	1.510	2.52
75	11983.277	83.856	78.239	5.617	−20.964	0.272	0.170	1.346	2.89
80	12530.185	92.667	89.873	2.794	−15.847	0.312	0.084	1.018	3.20
85	12897.622	98.526	97.778	0.748	−8.554	0.339	0.023	0.549	3.40
90	13027.500	100.587	100.587	0.000	0.000	0.349	0.000	0.000	3.47

Table 7.
Variation of tangential stress, material axis stress and tangential modulus at the edge of material discontinuity in PPSFC.

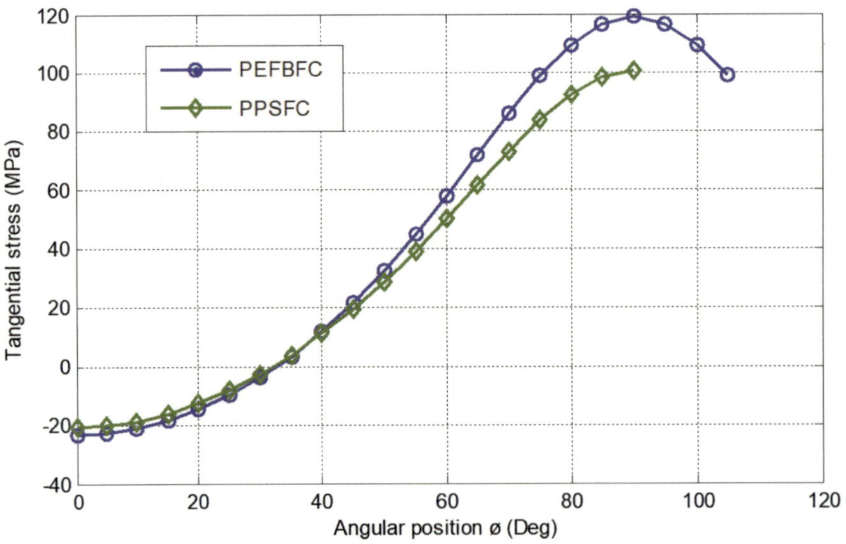

Figure 10.
Tangential stress distribution at a hole edge for PEFBFC and PPSFC.

fiber-reinforced composite as a result of structural discontinuity may not necessarily occur at the point of maximum stress concentration, therefore it is important to assess the extent of variation of the tangential stress around the hole edge and the failure index using maximum stress theory at other points aside the point of maximum stress concentration. Also in the present consideration we take a simplified scenario where $\sigma_2 = \sigma_{12} = 0$ such that the ply of dimensions $150 \times 19.05 \times 3.2$ mm with a circular hole at the center is subjected to only nominal axial stress $\sigma_{PEFBFC} =$ 34 MPa and $\sigma_{PPSFC} = 29$ MPa. **Tables 6** and 7 depict the values of tangential stress, material axis stress and tangential modulus as computed using Eq. (20)–(25).

Tangential stress distribution at hole edge for PEFBFC and PPSFC are shown in **Figure 10**, the maximum stress value of 119.15 and 100.587 MPa was attained at angular position ø = 90° for PEFBFC and PPSFC respectively. However, considering various failure indices in **Tables 5** and **6**, failure will be initiated at ø = 70° for PEFBFC with stress concentration factor of 2.53 and ø = 65° for PPSFC with stress concentration factor of 2.13 which are less than the stress concentration around the peak stress when angular position is 90°.

6. Conclusions

The utilization of plantain fiber-reinforced composites in structural applications empowers architects to acquire huge accomplishments in the usefulness, security and economy of development. These materials have high proportion of strength-to-density ratio, can be tailored to posses certain mechanical properties. The elastic constants of plantain fiber-reinforced composites depend greatly on fiber orientation with notable anisotropic characteristics which makes it less attractive for applications involving lugs and fittings. The present report amplified some notable design procedures in handling such limitations in plantain fiber-reinforced composites using relevant failure theories. Both plantain EFBFRC and PSFRC showed similar trends in response to the design scenario considered. Be that as it may be, a proficient utilization of plantain fiber-reinforced composites in structural applications requires a cautious assessment of all influential factors.

Author details

Christian Emeka Okafor[1*] and Christopher Chukwutoo Ihueze[2]

1 Department of Mechanical Engineering, Nnamdi Azikiwe University, Awka, Nigeria

2 Department of Industrial, Production Engineering, Nnamdi Azikiwe University, Awka, Nigeria

*Address all correspondence to: ce.okafor@unizik.edu.ng

References

[1] Dehmous H, Welemane H, Karama M, Tahar KA. Reliability approach for fibre-reinforced composites design. International Journal for Simulation and Multidisciplinary Design Optimization. 2008;**2**(1):1-9

[2] Okafor CE, Okafor EJ, Obodoeze JJ, Ihueze CC. Characteristics and reliability of polyurethane wood ash composites for packaging and containerisation applications. Journal of Materials Science Research and Reviews. 2018;**1**(3):1-10

[3] Xie HB, Wang YF. Reliability analysis of CFRP-strengthened RC bridges considering size effect of CFRP. Materials. 2019;**12**(14):2247

[4] Beaumont PW, Soutis C, Hodzic A, editors. The Structural Integrity of Carbon Fiber Composites: Fifty Years of Progress and Achievement of the Science, Development, and Applications. Switzerland: Springer; 2016

[5] Pei X, Han W, Ding G, Wang M, Tang Y. Temperature effects on structural integrity of fiber-reinforced polymer matrix composites: A review. Journal of Applied Polymer Science. 2019;**136**(45):48206

[6] Bittrich L, Spickenheuer A, Almeida JHS, Müller S, Kroll L, Heinrich G. Optimizing variable-axial fiber-reinforced composite laminates: The direct fiber path optimization concept. Mathematical Problems in Engineering. 2019;**2019**:1-11

[7] Prasad SV, Kumar GA, Sai KP, Nagarjuna B. Design and optimization of natural fibre reinforced epoxy composites for automobile application. In: AIP Conference Proceedings. Vol. 2128(1). Melville, NY, USA: AIP Publishing; 2019. p. 020016

[8] Wang Y, Cui G, Shao Z, Bao Y, Gao H. Optimization of the hot pressing process for preparing flax fiber/PE thermoplastic composite. Mechanical Engineering Science. 2019;**1**(1):41-45

[9] Foulc MP, Bergeret A, Ferry L, Ienny P, Crespy A. Study of hygrothermal ageing of glass fibre reinforced PET composites. Polymer Degradation and Stability. 2005;**89**(3):461-470

[10] Shettar M, Chaudhary A, Hussain Z, Kini UA, Sharma S. Hygrothermal studies on GFRP composites: A review. In: MATEC Web of Conferences. Vol. 144. Karnataka, India: EDP Sciences; 2018. p. 02026

[11] Ngo TD. Natural fibers for sustainable bio-composites. In: Natural and Artificial Fiber-Reinforced Composites as Renewable Sources. Rijeka: IntechOpen; 2018. p. 107

[12] Rahmani H, Najaf SHM, Ashori A, Golriz M. Elastic properties of carbon fibre-reinforced epoxy composites. Polymers and Polymer Composites. 2015;**23**(7):475-482

[13] Schürmann H. Konstruieren mit Faser-Kunststoff-Verbunden. Berlin Heidelberg: Springer-Verlag; 2007

[14] Younes R, Hallal A, Fardoun F, Chehade FH. Comparative review study on elastic properties modeling for unidirectional composite materials. In: Hu N, editor. Composites and their Properties. London: IntechOpen; 2012

[15] Goncalves R, Trinca AJ, Pellis BP. Elastic constants of wood determined by ultrasound using three geometries of specimens. Wood Science and Technology. 2014;**48**:269-287

[16] Longo R, Delaunay T, Laux D, El Mouridi M, Arnould O, Le Clézio E. Wood elastic characterization from a

single sample by resonant ultrasound spectroscopy. Ultrasonics. 2012;**52**: 971-974

[17] Majano-Majano A, Fernandez-Cabo JL, Hoheisel S, Klein MA. Test method for characterizing clear wood using a single specimen. Experimental Mechanics. 2012;**52**:1079-1096

[18] Vazquez C, Gonçalves R, Bertoldo C, Bano V, Vega A, Crespo J, et al. Determination of the mechanical properties of *Castanea sativa* mill. using ultrasonic wave propagation and comparison with static compression and bending methods. Wood Science and Technology. 2015;**49**:607-622

[19] Adewole S. Plantain (*Musa acuminata*) value chain analysis in Ondo State, Nigeria. Scientific Papers: Management, Economic Engineering in Agriculture & Rural Development. 2017;**17**(3):25-36

[20] Widsten P, Kandelbauer A. Adhesion improvement of lignocellulosic products by enzymatic pre-treatment. Biotechnology Advances. 2008;**26**:379

[21] Ojediran EO, Ibrahim HK, Adebisi LO, Belewu KY, Owolawi S. Analysis of profitability and determinants of plantain production in IFE agricultural development project (ADP) zone of Osun state, Nigeria. Fuoye Journal of Agriculture and Human Ecology. 2017;**1**(1):77-86

[22] HTA. Constraints to Banana Production. Ibadan, Nigeria: International Institute of Tropical Agriculture; 1990. pp. 142-144

[23] Norgrove L, Hauser S. Improving plantain (Musa spp. AAB) yields on smallholder farms in West and Central Africa. Food Security. 2014;**6**(4):501-514

[24] Akinyemi SOS, Aiyelaagbe IOO, Akyeampong E. Plantain (Musa spp.) cultivation in Nigeria: A review of its production, marketing and research in the last two decades. In: Dubois T, Hauser S, Staver C, Coyne D, editors. Proceedings of an International Conference on Banana & Plantain in Africa Harnessing International Partnerships to Increase Research Impact. Acta Horticulturae. Vol. 879. 2010. pp. 211-218

[25] Kaine AIN, Okoje LJD. Estimation of cost and return of plantain production in Orhionwon Local Government Area, Edo State, Nigeria. Asian Journal of Agriculture and Rural Development. 2014;**4**(2):162-168

[26] Kainga PE, Nnadi CD, Jimmy SP, Ugorji KS. Economics of plantain production in oil producing communities of Bayelsa State, Nigeria. Nigerian Journal of Agricultural Economics. 2015;**5**(1):54-60

[27] Ahmed S. The Survey of Nigeria Agriculture by Raw Materials Research and Development Council (RMRDC). 2004:123

[28] Olumba CC. Productivity of improved plantain technologies in Anambra State, Nigeria. African Journal of Agricultural Research. 2014;**9**(29): 2196-2204

[29] Samuel OD, Agbo S, Adekanye TA. Assessing mechanical properties of natural fiber reinforced composites for engineering applications. Journal of Minerals and Materials Characterization and Engineering. 2012;**11**(1):1-5

[30] Cadena Ch EM, Vélez R, M J, Santa JF, Otálvaro G. Natural fibers from plantain pseudostem (*Musa paradisiaca*) for use in fiber-reinforced composites. Journal of Natural Fibers. 2017;**14**(5):678-690

[31] Adeniyi AG, Ighalo JO, Onifade DV. Banana and plantain fiber-reinforced

polymer composites. Journal of Polymer Engineering. 2019;**39**(7):597-611

[32] Chimekwene CP, Fagbemi EA, Ayeke PO. Mechanical properties of plantain empty fruit bunch fiber reinforced epoxy composite. International Journal of Research in Engineering, IT and Social Sciences. 2012;**2**(6):86-94

[33] Okafor EC, Ihueze CC, Nwigbo SC. Optimization of hardness strengths response of plantain fibers reinforced polyester matrix composites (PFRP) applying Taguchi robust design. International Journal of Science & Emerging Technologies. 2013;**5**(1):1-11

[34] Alvarez-López C, Rojas OJ, Rojano B, Ganán P. Development of self-bonded fiberboards from fiber of leaf plantain: Effect of water and organic extractives removal. BioResources. 2014;**10**(1):672-683

[35] Obijiaku JC, Kamalu CIO, Osoka EC, Onyelucheya OE, Uzondu FN, Obibuenyi JI. Effects of extraction techniques on the yield and mechanical properties of empty plantain fruit bunch fibers. International Journal of Engineering and Management Research (IJEMR). 2015;**5**(6):494-500

[36] Ihueze CC, Okafor CE. Optimal design for flexural strength of plantain fibers reinforced polyester matrix. Journal of Innovative Research in Engineering and Sciences. 2016;**4**(4): 520-537

[37] Ihueze CC, Okafor EC. Response surface optimization of the impact strength of plantain fiber reinforced polyester for application in auto body works. Journal of Innovative Research in Engineering and Science. 2014;**4**:505-520

[38] Okafor CE, Godwin HC. Evaluation of compressive and energy adsorption characteristics of plantain fiber reinforced composites. World Journal of Engineering and Physical Sciences. 2014;**2**(3):036-048

[39] Oreko BU, Otanocha OB, Emagbere E, Ihueze CC. Analysis and application of natural fiber reinforced polyester composites to automobile fender. Covenant Journal of Engineering Technology (Special Edition). 2018;**1**(1): 1-12

[40] Imoisili PE, Tonye DI, Victor PA, Elvis OA. Effect of high-frequency microwave radiation on the mechanical properties of plantain (*Musa paradisiaca*) fibre/epoxy biocomposite. Journal of Physical Science. 2018;**29**(3):23-35

[41] Imoisili PE, Fadare OB, Popoola AV, Okoronkwo AE. Effect of chemical treatment on the morphology and mechanical properties of plantain (*Musa paradisiaca*) fibre. IOSR Journal of Applied Chemistry. 2017;**10**(5):70-73. DOI: 10.9790/5736-1005017073

[42] Sinebe JE, Chukwuneke JL, Omenyi SN. Implications of interfacial energetics on mechanical strength of fiber reinforced polymer matrix. International Journal of Materials Engineering. 2019;**9**(1):1-7. DOI: 10.5923/j.ijme.20190901.01

[43] Ihueze CC, Okafor CE, Okoye CI. Natural fiber composite design and characterization for limit stress prediction in multiaxial stress state. Journal of King Saud University-Engineering Sciences. 2015;**27**(2):193-206

[44] Kenedi PP, Vignoli LL, Duarte BT, Matos FCDA, Dias HOT. Orthotropic elastic properties assessment of sandwich laminates. Journal of Aerospace Technology and Management. 2017;**9**(3):389-396

[45] Hwang SF, Liu HT. Prediction of elastic constants of carbon fabric/ polyurethane composites. In: Solid State Phenomena. Vol. 258. Baech,

Switzerland: Trans Tech Publications; 2017. pp. 233-236

[46] Ren B, Noda J, Goda K. Effects of fiber orientation angles and fluctuation on the stiffness and strength of sliver-based green composites. Journal of the Society of Materials Science, Japan. 2010;**59**(7):567-574

[47] Kumar KV, Reddy PR, Shankar DR. Influence of angle ply orientation of stacking on mechanical properties of glass-polyester composite laminate. International Journal of Engineering and Advanced Technology (IJEAT). 2013; **2249**:8958

[48] Cordin M, Bechtold T, Pham T. Effect of fibre orientation on the mechanical properties of polypropylene–lyocell composites. Cellulose. 2018;**25**(12):7197-7210

[49] Ihueze CC, Okafor EC, Ujam AJ. Optimization of tensile strengths response of plantain fibers reinforced polyester composites (PFRP) applying Taguchi robust design. Innovative Systems Design and Engineering. 2012; **3**(7):64-76

[50] Jules EJ, Tsujikami T, Lomov SV, Verpoest I. Effect of fibres length and fibres orientation on the predicted elastic properties of long fibre composites. Macromolecular Symposia. 2004;**17**:1-109

[51] Venetis J, Sideridis E. Elastic constants of fibrous polymer composite materials reinforced with transversely isotropic fibers. AIP Advances. 2015; **5**(3):037118

[52] Cuartas VM, Perrin M, Pastor ML, Welemane H, Cantarel A, Karama M. Determination of the elastic properties in CFRP composites: Comparison of different approaches based on tensile tests and ultrasonic characterization. Advances in Aircraft and Spacecraft Science. 2015;**2**(n° 3):249-260

[53] Clyne TW, Hull D. An Introduction to Composite Materials. Cambridge, United Kingdom: Cambridge University Press; 2019

[54] Datoo MH. Mechanics of Fibrous Composites. New York: Elsevier Science Publishing Co. Inc; 2012

[55] Maheshwari P, Misra A, Kumar A. Investigate the variation in elastic constants and failure strength of lamina with FVF and FOA for JFRP lamina. International Journal of Applied Engineering Research. 2018;**13**(9): 157-161

[56] Jones RM. Mechanics of Composite Materials. 2nd ed. Philadelphia, PA: Taylor and Francis, Inc.; 1999

[57] Gibson RF. Principles of Composite Material Mechanics. 4th ed. Boca Raton, FL: CRC Press; 2016

[58] Farooq U, Myler P. Efficient determination of mechanical properties of carbon fibre-reinforced laminated composite panels. ARPN Journal of Engineering and Applied Sciences. 2017; **12**(5):1375-1392

[59] Koussios S, Beukers A. Lekhnitskii's formalism for stress concentrations around irregularities in anisotropic plates: Solutions for arbitrary boundary conditions. In: Variational Analysis and Aerospace Engineering. New York, NY: Springer; 2009. pp. 243-265

[60] Lekhnitskii SG. Anisotropic Plates. 2nd ed. London, UK: Gordon and Breach; 1968

Chapter 12

Surface Measurement and Evaluation of Fiber Woven Composites

Bin Lin, Haoji Wang and Jinhua Wei

Abstract

The surfaces of fiber woven composites (FWCs), especially woven ceramic matrix composites (WCMCs), are obviously anisotropic. Many kinds of damage, which are different from traditional homogeneous materials, could be caused by the fabrication and machining process. The old surface evaluation system appropriate for isotropic materials is no longer suitable to WCMCs, thus causing many difficulties in terms of their wide industrial applications. This chapter presents a grading surface measurement and evaluation system for WCMCs based on their microstructures. The system includes four levels: fiber, fiber bundle, cell body, and the whole surface. On the fiber level, the typical forms of fiber damage, and their effects on the surface morphology of WCMCs are analyzed, which lays a foundation for the measurement and evaluation methods on the next three levels. On each subsequent level, the system proposes a set of surface measurement sampling parameter determination methods and surface quality evaluation methods based on the principle of statistics. As demonstrations, the surface measurement and evaluation on each level were processed on a carbon fiber-reinforced silicon carbide matrix composite (C_f/SiC) to illustrate the methodology of the system.

Keywords: woven ceramic matrix composites, grading surface evaluation, surface sampling method, confidence interval, residual error estimate

1. Introduction

Fiber woven composites (FWCs) are a kind of new fiber composites. The fibers inside are woven to form a preform, and then the matrix grows on the preform to generate the final composites. The woven fibers can heighten the reinforcing effect of the fibers and improve the mechanical property of the composite. Therefore, FWCs could perform better than other fiber composites.

Among FWCs, woven ceramic matrix composites (WCMCs) are star materials, widely used in aerospace, military, national defense, and some other advanced fields [1–4], because of their high specific strength and rigidity, corrosion and wear resistance, and other excellent characteristics [5, 6]. For the industrial application of WCMCs, it is of vital importance to objectively evaluate the surface processing quality and, on this basis, judge the type and degree of processing damage. To do so, it is essential to measure the surface topography both accurately and efficiently and select proper indexes to evaluate the surface process quality. However, WCMCs are far more complicated than traditional materials. On the one hand, their surfaces are anisotropic and inhomogeneous and have obvious directionality and complex structures, which means that there are difficulties and challenges in measuring and evaluating their surfaces. Traditional surface measurement and evaluation approaches for isotropic materials are no longer suitable to WCMCs [7–9]. On the other hand, the surfaces of an WCMC present more types of processing damage than isotropic materials, including fiber pullout, debonding, and matrix cracking [10]. Each shows a different influence on the composite application, and thus judging the type and degree of processing damage to a WCMC is a new but difficult task.

To date, there have been no uniform measurement standards to ensure that undistorted WCMC surface features are obtained or proper evaluation approaches accurately assess the surface damage [11, 12]. It is widely believed that only 3D measurements can obtain the complete surface information [7, 13, 14]. However, a traditional evaluation method used to assess the isotropic materials is limited to a quantitative description of the entire surface through some typical surface topography parameters, which ignores the subtle details of the surface. Such judgment standards are brief with respect to the direct relation to the surface damage. Moreover, to date, a majority of composite surface evaluations still use the profile arithmetic mean error *Ra* as the only evaluation parameter [15–20], which is fairly incomplete.

As such, the complexity of a WCMC surface calls for a newer and more targeted methodology that is tightly connected with the topography characteristic. When we look into the WCMC surface, it is obvious that its composition sequence is as follows: fiber -fiber bundle -cell body -whole surface [21]. Here, a fiber is the smallest composing unit, a fiber bundle is the smallest structural unit, and a cell body is the smallest repeatable unit. A cell body is made up of fiber bundles and matrix and has a nearly fixed surface microstructure. The material surface is formed through its repeating copy and translation [22, 23]. Thus, fiber damage influences the fiber bundle surface, damage to the fiber bundle surface influences the cell body surface, and damage to the cell body surface influences the whole surface property. Merely depicting the entire surface at one time without considering the surface structure composition of a WCMC is inadvisable.

In this chapter, it is proposed that the measurement and evaluation of a WCMC surface should adopt a grading evaluation system based on its complex surface structure, which includes the four levels: fiber, fiber bundle, cell body, and the whole surface. On the fiber level, the typical forms of fiber damage and their effects on the surface morphology of WCMCs are analyzed, which lays a foundation for the measurement and evaluation methods on the next three levels. On each subsequent level, the system proposes a set of surface measurement sampling parameter determination methods and surface quality evaluation methods based on the principle of statistics.

2. Fundamental concepts, devices, and materials of the research

2.1 The microstructure of a WCMC surface

According to the common fabrication process of WCMCs (shown in **Figure 1**), single fibers are surface modified to improve the bonding strength between fibers and matrix, and then several fibers are twined to form a fiber bundle. Multiple fiber bundles are woven in a certain way to a preform. The preform is then immersed

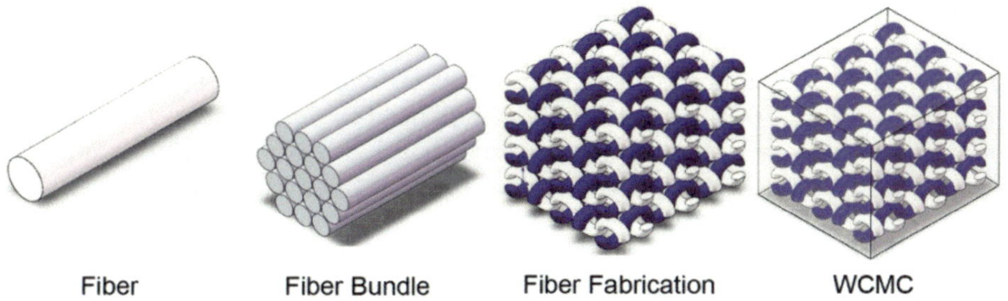

Figure 1.
The fabrication process of an WCMC.

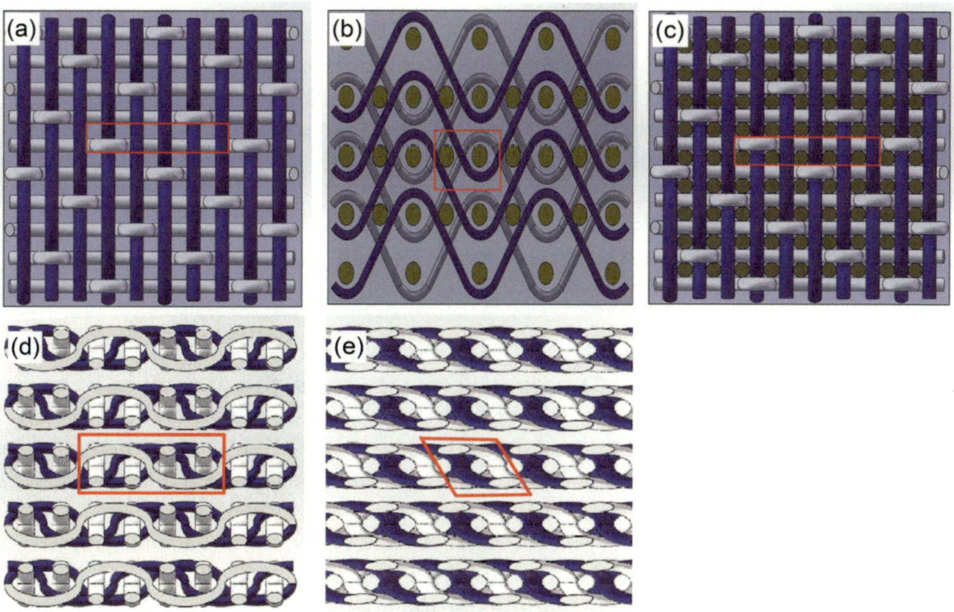

Figure 2.
Schematic diagram of different woven patterns of WCMC. (a) 2D, (b) 2.5D, (c) 3D, and (d) 2D woven style with a processing angle of 90° and (e) 2D woven style with a processing angle of 45°.

into an environment with the elements or components of the matrix. The matrix can grow on the preform to generate the final WCMCs.

The schematics of the WCMC surfaces of different woven methods (shown in **Figure 2(a–c)**) and process angles (shown in **Figure 2(d, e)**) indicate that there exists a minimum repeatable unit, which is marked with a red block in the figures. The unit is composed of the fiber bundles of every directions and the ceramic matrix. The whole surface can be formed through its repeating copy and translation. The unit is defined as the "cell body" in this chapter. It is obvious that the shape of a cell body is not uniform for a WCMC. In fact, the appearance of a cell body can change with different woven methods and process angles.

In summary, fiber is the minimum characteristic of a WCMC, which forms fiber bundle. Fiber bundles are woven to different directions. The cell body consists of fiber bundles of every directions and the matrix. Eventually, the whole surface is generated by copy and translation of the cell body. Therefore, fiber is the minimum evaluable unit of WCMCs. Its damage form can influence the surface state of fiber bundles. Fiber bundles build a bridge between the "microscale" of the fiber and the "macroscale" of the cell body. Their surfaces include the information of the fibers and impact the surface quality of the cell body. Cell body is a key feature. On the one hand, its components are as complicated as the whole surface, which means that the evaluation results of a cell body can be used to represent and estimate a certain range of a whole surface. On the other hand, it is obviously affected by the fiber bundles and matrix inside; thus the analysis of fiber and fiber bundle can be used to evaluate the cell body.

In this chapter, it is believed that the measurement and evaluation of WCMCs should employ a grading system. The evaluation of fiber, fiber bundle, cell body, and the whole surface should be separately researched and then integrated. On the fiber level, the typical damage forms of fibers should be identified and classified, and how the damage influence the using properties should be research. On the subsequent levels, two tasks should be accomplished. The first one is to select proper sampling methods to acquire the undistorted surface information. And the second one is to propose and test reliable evaluation indexes to quantitatively estimate the main damage type and degree of the surfaces. With all the work above, a grading measurement and evaluation system for the surface of a WCMC can be eventually built. The following parts of this chapter introduce the methodologies on each level.

2.2 The effect of fiber damage on the surface measurement and evaluation of WCMCs

When processed by machining tools, the surfaces of WCMCs interact with the cutting edges, leading to fiber damage. The damage of fiber on the one hand causes the removal of fibers and, on the other hand, turns into the machining defects on the surfaces. The multifarious types of the damage of WCMCs are the main feature of difference to the traditional homogeneous materials and, meanwhile, are the main source of technological difficulty of the evaluation of WCMCs.

The typical forms of damage on the surfaces of WCMCs are the following:

Fiber fracture, which is caused by the cutting edges directly cutting the fibers off, often happens when the fibers at the cutting area are tightly fixed by the matrix or by other fibers nearby. The cutting section of fiber fracture is V-shape, and the bottom of the V often appears plastic deformation, which is caused by the friction and squeezing between the fibers and the cutting edges. Fiber fracture is the most common material removal form of WCMCs. If the materials are mainly removed by fiber fracture, the finished surfaces are always of good quality and with low roughness.

Fiber pulled-off is caused by the cutting edges breaking the matrix without cutting off the fibers or the cutting edges pulling the fibers out of matrix without cutting them off. This type of damage can leave fibers exposed or form holes on the finished surface, which decreases the surface quality. When the finished surface is assembled with another part, the raised fiber can act as a tiny cutting edge, harming the counterpart surface. When the finished surface performs as a friction surface, the holes may help contain lubricating oil and wear debris, alleviating three-body wear, thus improving its tribological performances.

Fiber debonding, which is caused by the cutting edges removing the entire layer of fibers, often occurs when the fibers at cutting area are poorly connected with the matrix. Fiber debonding can result in collapses of large areas on the surface and greatly reduce the surface quality.

From the analysis above, it is clear that the machining process can cause manifold types of damage on the surface of WCMCs. Each kind of damage can affect the quality and performance of the surface in its own way. Therefore, the evaluation technology of WCMCs is required to recognize the main type of damage on a surface and quantitatively estimate the degree of the damage.

Moreover, fiber bundles, cell bodies, and the whole surface are, basically speaking, made up of fibers. Because of the directional arrangements of fibers, the height of the surface of WCMCs could fluctuate with the period of fiber diameter. The fiber diameter acts as an obvious fundamental frequency on the surface of WCMCs. It can influence the sampling parameters, such as sampling step, sampling length, and sampling area, on other levels. Meanwhile, the direction of fibers determines the direction of the fundamental frequency and can eventually influence the sampling direction. Thus, the direction and diameter of fibers are tightly connected with the surface measurement technology of WCMCs.

In conclusion, fiber, as the minimum evaluable unit of a WCMC surface, significantly influences the grading surface measurement and evaluation system of WCMCs. On the one hand, when considering whether an evaluation index is appropriate for WCMC surface, it should be checked whether this index can help recognize and estimate the type and degree of damage. On the other hand, when determining the proper sampling parameters, the directionality and the diameter of fibers must be taken into consideration.

2.3 Surface measurement technology and devices

The measurement of a surface is a process that obtains the height information of the surface. Surface measurement methods are divided into two classes: contact measurement, which uses a probe to measure the height data of points on the surface, and non-contact measurement, which uses light to measure. It was always believed that contact measurement could achieve higher measurement accuracy, although its efficiency was quite low and the measurement process was time-consuming. However, thanks to the development of optical theories and technologies, the non-contact measurement technology based on white light interferometry can get extremely high level of accuracy now as well. NANOVEA ST400 (shown in **Figure 3**), an optical non-contact measurement system, is used to measure the surface micro-topography in the research of this chapter.

Because the measurement of a surface is in fact the measurement of the points on the surface, it has to be determined which points are chosen to be measured. This

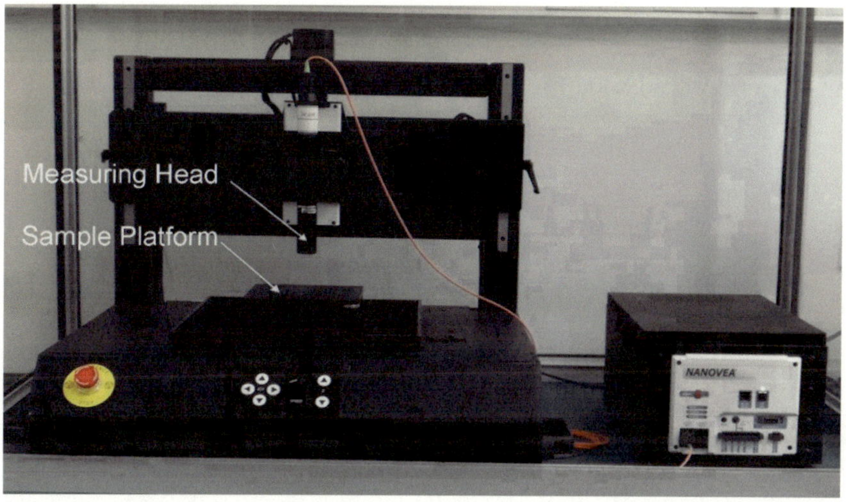

Figure 3.
Three-dimensional non-contact surface morphometer.

topic is related to the sampling strategy, which means how to choose a proper set of points to measure, in order to make the measurement results of these samples able to reflect the information of the entire surface. The first question is to use 2D measurement or 3D measurement.

If a 2D measurement is adopted, points in a line are measured and calculated as one data set. In this term, sampling step (the length between two adjacent sampled points), sampling length (the length of the entire sampling line), and sampling direction (the angle between the sampling line and the structure of the surface) should be determined. In most instances of 2D measurement, one line of sampling is not able to reflect the entire surface because of the measuring error and the random surface damage. Several lines should be selected and measured in order to improve the stability of the measurement results. Therefore, sampling number (the number of the sampling lines) is also to be determined.

If a 3D measurement is adopted, an array of points inside the entire surface are measured and calculated as one data set. That's to say, the sampling area is the area of the entire surface. When sampling step is determined, the points to be measured are selected. It was believed that 3D measurement was more adaptable for complex surfaces because it could get more information of the surfaces. However, the research of this chapter proved that, if the surface to be measured is obviously directional, 3D measurement will lose its advantages and 2D measurement should be adopted. On the other side, 3D measurement always means long sampling time, huge data processing work, and, thus, low efficiency.

Selecting proper sampling parameters, including sampling step, length, direction, number of 2D measurement, and sampling step of 3D measurement, is a complex work. Small step and large length and number are always related to higher measuring accuracy but low efficiency and vice versa. Proper sampling parameters balance both two sides, maintain undistorted sampling, and, on this basis, reduce sampling points.

2.4 Surface evaluation technology

The task of surface evaluation technology is to select proper statistical characteristics (defined as evaluation indexes in this chapter), which can be calculated from the measurement data of the heights of the sampling points. The following are the indexes adopted in the research of the chapter:

For 2D measurement data:

For each sampling profile j, its average is defined as μ_{0j}, the standard deviation is σ_{0j}, and the normalized height of every sampling point is Z_{0ij}, and thus

$$\mu_{0j} = \sum_{i=1}^{M} Z_{ij}/M \tag{1}$$

$$\sigma_{0j} = \sqrt{\sum_{i=1}^{M} \left(Z_{ij} - \mu_{0j}\right)^2/(M-1)} \tag{2}$$

$$Z_{0ij} = \left(Z_{ij} - \mu_{0j}\right)/\sigma_{0j} \tag{3}$$

where the height of every sampling point is defined as Z_{ij}, j is the jth profile on the fiber bundle surface, and the number of sampling points within the sampling profile is M, where M = the *sampling length/sampling step*.

Based on the above, the four 2D evaluation indexes used in the chapter are profile arithmetic mean error *Ra*, profile square root deviation *Rq*, profile skewness *Rsk*, and profile kurtosis *Rku*, which can be calculated as follows:

$$Ra_j = \sum_{i=1}^{M} |Z_{ij}|/M \tag{4}$$

$$Rq_j = \sqrt{\sum_{i=1}^{M} Z_{ij}{}^2/M} \tag{5}$$

Here, *Ra* and *Rq* are quite similar in reflecting the surface roughness, although *Rq* is in general more sensitive than *Ra* to the degree of surface roughness.

The normalized R_{sk0} and R_{ku0} of the *j*th profile can be obtained as follows:

$$Rq_{0j} = \sqrt{\sum_{i=1}^{M} Z_{0ij}{}^2/M} \tag{6}$$

$$Rsk_{0j} = \sum_{i=1}^{M} Z_{0ij}{}^3/\left(MRq_{0j}{}^3\right) \tag{7}$$

$$Rku_{0j} = \sum_{i=1}^{M} Z_{0ij}{}^4/\left(MRq_{0j}{}^4\right) \tag{8}$$

According to the definition of *Rsk*, the closer it is to 0, the more approximate the sampling profile is to a Gaussian distribution. When $Rsk > 0$, the profile presents a positive peak. This indicates that the profile has more crests or the crest height is larger than the trough height. In contrast, if a profile has more troughs, or the trough height is larger than the crest height, it presents a negative peak.

On the other hand, *Rku* is compared with 3. The closer it is to 3, the more approximate the sampling profile is to a Gaussian distribution. That is, the degree of dispersion of the profile data is similar to a Gaussian distribution profile. The more *Rku* is greater than 3, the smaller the degree of data dispersion is, and in contrast, the more *Rku* is less than 3, the larger the degree of data dispersion is.

For 3D measurement data:

The four 3D evaluation indexes used in the chapter are surface arithmetic mean deviation *Sa*, surface square root deviation *Sq*, surface skewness *Ssk*, and surface kurtosis *Sku*, which can be calculated as follows:

$$Sa = \sum_{i=1}^{M}\sum_{j=1}^{N} |Z_{ij}|/M/\mathrm{N} \tag{9}$$

$$Sq = \sqrt{\sum_{i=1}^{M}\sum_{j=1}^{N} Z_{ij}{}^2/M/\mathrm{N}} \tag{10}$$

$$Ssk = \sum_{i=1}^{M}\sum_{j=1}^{N} Z_{ij}{}^3/\left(MNSq^3\right) \tag{11}$$

$$Sku = \sum_{i=1}^{M}\sum_{j=1}^{N} Z_{ij}{}^4/\left(MNSq^4\right) \tag{12}$$

where the height of every sampling point is defined as Z_{ij} and the number of sampling points within the sampling area is *M* and *N*, where *M* and *N* = the *sampling length* (of X and Y direction, respectively)/*sampling step*.

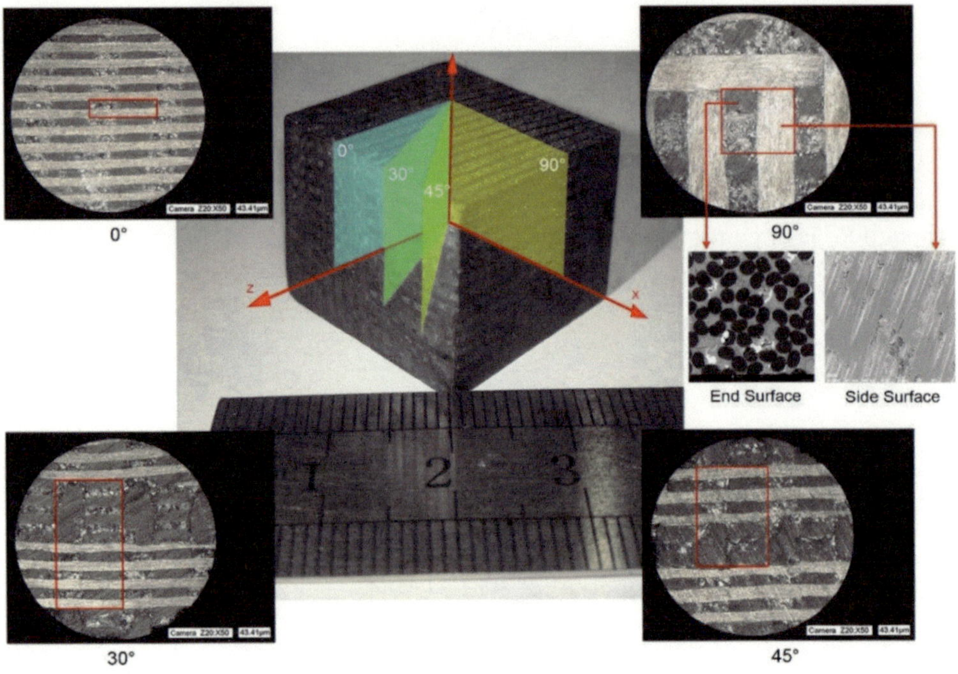

Figure 4.
Definition of the processing angle of C_f/SiC [26].

2.5 Materials used as examples in the chapter

In order to illustrate the measurement and evaluation method, several materials were measured and evaluated as examples in this chapter. The information of the materials is shown as follows:

The carbon fiber-reinforced silicon carbide ceramic matrix composite (C_f/SiC) was fabricated through chemical vapor infiltration (CVI) combined with a liquid melt infiltration process (LMI) [24]. The preform was prepared using a 3D needling method and densified using CVI to form a porous carbon/carbon (C/C) composite. Next, the porous C/C composite was converted into C_f/SiC during LMI, in which silicon carbide (SiC) matrix was formed through a reaction with carbon and melted silicon [25]. The density of the C_f/SiC composite is 1.85 g/cm^3.

The fiber diameter of the material is about 7 μm, and size of the cell body is about 1.6 mm × 1.6 mm.

The C_f/SiC specimens were ground with four different processing angles. For 90° processing angle, the fiber bundles are divided into side fiber bundles and end fiber bundles, according to their directions (shown in **Figure 4**).

3. The measurement and evaluation of fiber bundle surfaces

3.1 Measurement strategy and sampling direction

When measuring fiber bundle surfaces, 2D measurement should be adopted because of the obvious directionality of the surfaces. Here we take the surface of C_f/SiC with a processing angle of 90° as an example. Owing to the directionality of the fiber bundles, different sampling directions often result in different numerical characteristics, as shown in **Figures 5** and **6**. We can see that no matter what type of measurement direction is applied, there is no influence on the 3D surface topography. That is, a 3D sampling and evaluation method may not be able to reflect the surface details of the fiber bundle. The use of one or a group of 3D evaluation indexes based on a 3D sampled data fails to reflect the damage types related to the fiber orientation and machining direction.

On the side surface of a fiber bundle, the bonding strength between the fiber and matrix is weaker than that of the end surface. The most direct reflection of the machining direction is fiber damage such as fiber debonding, fiber fractures and delamination. The fiber direction scale is more notable than the machining direction scale, and the directionality of the surface topography mainly depends on the fiber orientation. On the end surface of a fiber bundle, the fiber is mainly subjected to a shear force. The main fiber damage is fiber shearing and fiber pullout. The machining

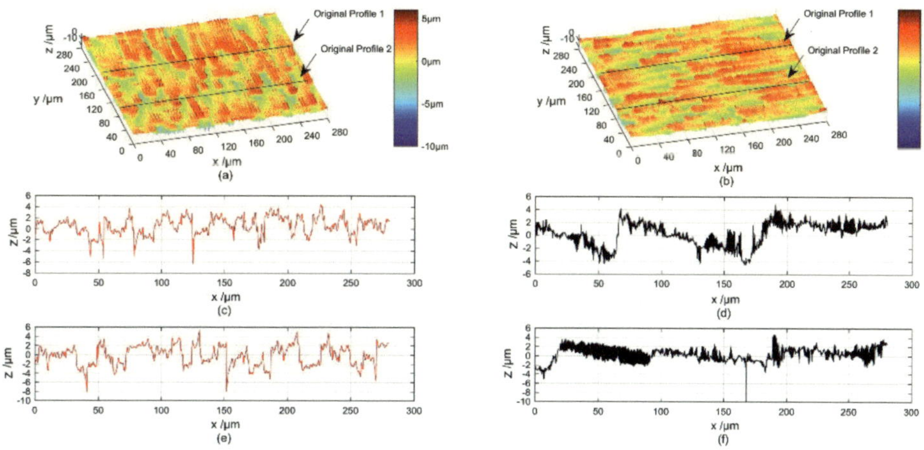

Figure 5.
Side surface topography of a fiber bundle with a scanning track perpendicular and parallel to the fiber direction [27]. (a) Surface topography-scanning track perpendicular to the fiber direction. (b) Surface topography-scanning track parallel to the fiber direction. (c) Original profile 1, scanning track perpendicular to the fiber direction. (d) Original profile 1, scanning track parallel to the fiber direction. (e) Original profile 2, scanning track perpendicular to the fiber direction. (f) Original profile 2, scanning track parallel to the fiber direction.

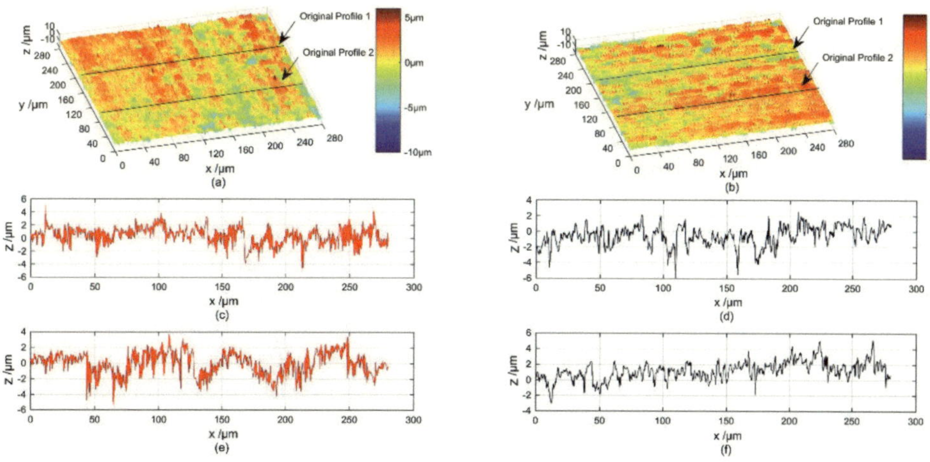

Figure 6.
End surface topography of a fiber bundle with a scanning track perpendicular and parallel to the machining direction [27]. (a) Surface topography-scanning track perpendicular to the machining direction. (b) Surface topography-scanning track parallel to the machining direction. (c) Original profile 1, scanning track perpendicular to the machining direction. (d) Original profile 1, scanning track parallel to the machining direction. (e) Original profile 2, scanning track perpendicular to the machining direction. (f) Original profile 2, scanning track parallel to the machining direction.

direction scale is more notable than the fiber orientation scale, and the directionality of the surface topography mainly depends on the machining direction.

On the side surface of a fiber bundle, as shown in **Figure 5(c, e)**, when the scanning track is perpendicular to the fiber direction, the profile shows damage between fibers, whereas the profiles only show single fiber damage when the scanning track is parallel to the fiber direction (**Figure 5(d, f)**), which means that the profiles cannot reflect the machining effect on the whole fiber bundle surface. The same phenomenon occurs in the end surface of a fiber bundle. When the scanning track is perpendicular to the machining direction, the profile shows the integrated influence of the processing (**Figure 6(c, e)**); however, when the scanning track is parallel to the machining direction, the profiles simply show the effect of a single grain on the surface (**Figure 6(d, f)**).

From the analysis above, it can be seen that the 2D sampling and evaluation method is more suitable for a fiber bundle scale measurement. To guarantee measurement accuracy and consider the influence of the fiber orientation and machining direction on the surface topography, the scanning track should be perpendicular to the fiber orientation on the side surface of a fiber bundle and perpendicular to the machining direction on the end surface.

According to our research, it may be reasonably inferred that, for planes which are not truly along the fibers, the influence of the fiber orientation and machining direction should be considered. When the surface is full of processing traces and the fiber orientation is so obscure, the sampling direction should be perpendicular to the machining direction. In other cases, the sampling direction should still be perpendicular to the projection direction on the vertical plane along the fiber axis. However, a definite conclusion in this regard still requires further research.

3.2 Determination of sampling length and number

The side surface of C_f/SiC with processing angle of 90° is taken as an example to illustrate the determination method of sampling length when measuring a fiber bundle surface. Since the diameters of the fibers are approximately 7 μm, a set of candidate sampling length are chosen as the integral multiple of 7 μm, namely, 7, 14, 21, 28, 35, 42, 49, 56, 63, 70, 77, 84, 91, 98, 105, 112, 119, 126, 133, 140, and 147 μm. In each sampling length, 1500 surface profiles are measured with a constant sampling step of 0.1 μm. And 2D surface roughness Ra of each profile is obtained.

According to the numerical values of 1500 Ra, a frequency histogram is made. It can be seen that the distribution of 2D surface roughness Ra in every sampling length is almost of its normal distribution. The result shown in **Figure 7** is the one using normal distribution function to fit the frequency histogram. With the growth of sampling length (**Figure 8(a)**), the curves are thinner and higher. Their shapes do not change any more in the case that sampling length is more than a certain value (**Figure 8(b)**). It is known to all that normal distribution has two parameters: the mean value μ and the standard deviation σ. μ is the location parameter and describes the central tendency position of the normal distribution. σ demonstrates the discrete degree of data. The larger the σ is, the more decentralized the data is, leading to a fact that the curve is fatter and lower. On the contrary, the more concentrated the data is, the thinner and taller is the curve. That is to say, Ra is gradually convergent and concentrated while the sampling length increases.

Figure 8 clearly shows the changing trends of standard deviation σ and the mean value Ra_{ave} of 1500 2D surface roughness Ra under different sampling length. It can be found that with the increase of sampling length, both σ and Ra_{ave} are gradually

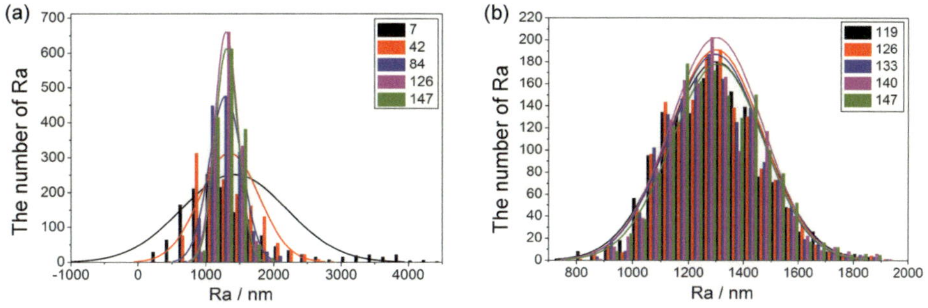

Figure 7.
The distribution of 2D surface roughness Ra under different sampling length [24]. (a) With the sampling lengths of 7,42 84 125 and 147. (b) With the sampling lengths of 119,126 133 140 and 147.

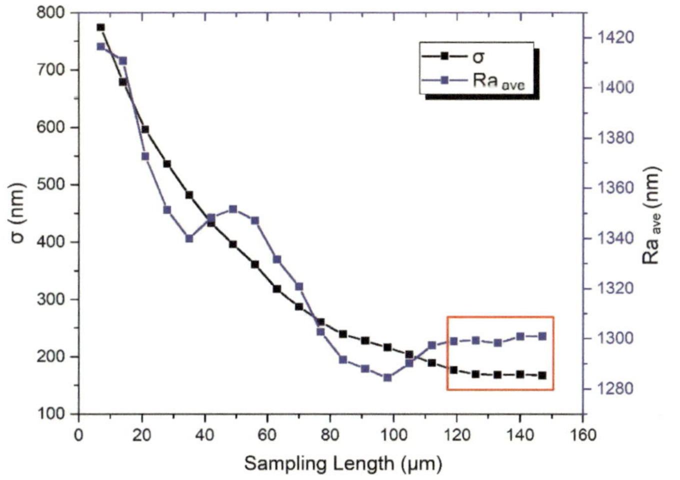

Figure 8.
The changing trends of σ and Ra_{ave} under different sampling length [24].

decreasing and becoming steady. When the sampling length reaches to 120 μm, Ra_{ave} is stable around 1297.3 nm.

Based on the results obtained above, a speculation is proposed that the mean value of a few number of *Ra* can steadily estimate the entire surface roughness of fiber bundle. The average value of *Ra* under the changing sampling number from 10 to 150 and the constant sampling length 120 μm could be obtained, which is represented by Ra_{avg}. Under every sampling number, the measurement process repeats 50 times independently. Then the maximum relative error of every sampling number, calculated by Eq. (13), is demonstrated in **Figure 9**. It is shown that with the rise of sampling numbers, the maximum relative error would decrease dramatically. Once the number of *Ra* reaches to 70 or above, it is stable under 2%, which is acceptable in terms of accuracy.

$$\delta = \left|Ra_{ave} - Ra_{avg}\right| / Ra_{ave} * 100\% \tag{13}$$

According to the analysis given above, the conclusion can be made that as long as extracting surface profiles averagely distributed on the side surface of C_f/SiC composite fiber bundle with the appropriate sampling length and sampling number, the mean value of Ra is steady and can estimate the whole surface roughness. For

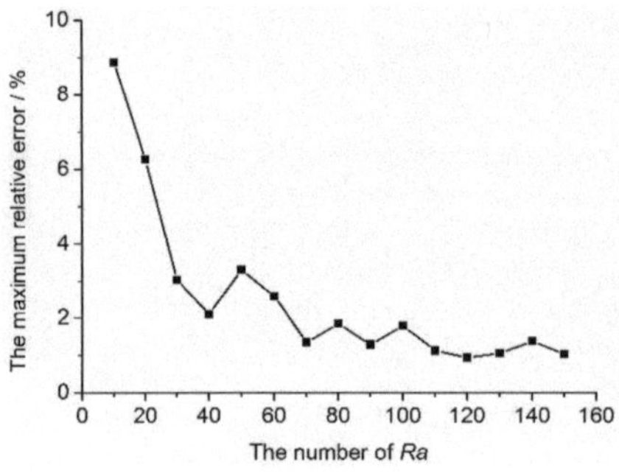

Figure 9.
The changing trend of the maximum relative error under different number of Ra [24].

C_f/SiC composite used in the present work, the critical sampling length is 119 μm, which is about 17 times of the fiber diameter, and sampling number is 70.

3.3 Determination of sampling step

In this section, the side surface and end surface of C_f/SiC with processing angle of 90° are taken as an example to illustrate the determination method of sampling step, under the condition that the critical sampling length is 150 μm and sampling number is 200. It is clear that using a smaller sampling step can achieve more accurate surface data, whereas a too small step may cause an unnecessary sampling time and data processing cost. A method is proposed to determine the maximum sampling step (MaxSS) that can minimize the data size under the premise of undistorted surface sampling.

We start from setting the data measured at the step of 0.05 μm as a certain type of real value μ of surface topography parameters. The data measured using larger steps are to be compared with the real values μ to determine whether they are acceptable in terms of accuracy. To set a range of acceptance, the idea of a confidence interval in probability theory is used. If the real value is μ, a measurement result that is acceptable based on confidence level of 1-α must fall into a computable interval. Based on the probability theory, when the mean value of the overall sample μ is known and the standard deviation σ is unknown, the confidence interval of the mean value μ with the confidence level $(1-\alpha)$ is

$$\left[\mu - t_{\alpha/2}(n-1)S/\sqrt{n}, \mu + t_{\alpha/2}(n-1)S/\sqrt{n}\right] \tag{14}$$

where S is the standard deviation of the samples and n is the number of the samples.

By looking up the table-α quantile of the t-distribution, $t_{\alpha/2}(n-1)$ is available, and thus the corresponding confidence intervals are obtained. Therefore, the sampling step gradually increases until the measurement result falls out of the acceptance range at that step. This means that this step, and the steps larger than it, can no longer achieve accurate surface data. The largest permitted sampling step can be determined under each single evaluation index. Combining all indexes, the global MaxSS can be determined.

Here, Ra_{ave}, Rq_{ave}, Rsk_{ave}, and Rku_{ave} on a Cf/SiC fiber bundle surface are taken as the evaluation standards. For each index, the real value is determined as the value measured based on a sampling length of 150 μm, sampling number of 200, and sampling step of 0.05 μm. In addition, $t_{\alpha/2}(n-1)$ can be found to be 2.326 when the confidence coefficient is 98%. After that, μ and S can easily be calculated. The acceptance range of each index is obtained through Eq. (14). The sampling length and sampling number are invariable, and the sampling step is gradually increased. All measurement results are shown in **Figure 10** (side surface) and **Figure 11** (end surface).

What needs to be mentioned is that the data used in **Figures 10** and **11** are acquired using a sampling direction perpendicular to the fiber orientation on the side surface and the machining direction on the end surface. For each index, the upper and lower limits express the acceptance range. The measurement result that first falls out of range is marked with a red box, and the last sampling step before it is the MaxSS of this index. Combining all four indexes on each surface, the global MaxSS is 0.75 μm on a side surface and 1 μm on an end surface, which are approximately 1/10 of the fiber diameter.

3.4 Evaluation indexes

This chapter proposes four indexes for evaluating a fiber bundle surface, namely, *Ra*, *Rq*, *Rsk*, and *Rku*. To illustrate how the four indexes can estimate the main type and degree of damage of a fiber bundle surface, the side surface and end surface of C_f/SiC with processing angle of 90° are taken as a demonstration. The surfaces of C_f/SiC were process with three machining methods: a) ground using a grinding wheel with a wheel speed of 15 m/s, grinding depth of 0.15 mm, feed rate of 4 m/min, and grain mesh size of 80#; b) polished using a 1200# sandpaper under a constant force of 5 N, spindle speed of 0.1 m/s, and sliding time of 60s; and c) friction against a ZrO_2 disk under a constant force of 30 N, spindle speed of 0.5 m/s, and sliding time of 3600 s. The three different methods caused different surface topographies and damages. Therefore, a proper set of indexes should be able to reflect the difference of the six kinds of surfaces.

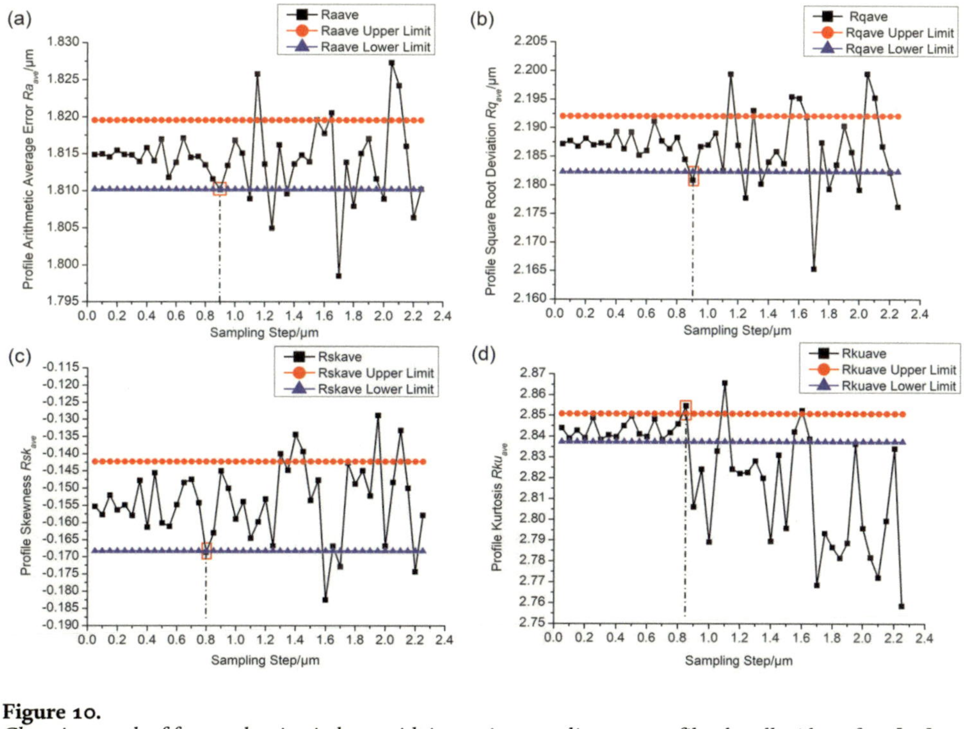

Figure 10.
Changing trends of four evaluation indexes with increasing sampling steps on fiber bundle side surface [27]. (a) Ra_{ave}, (b) Rq_{ave}, (c) Rsk_{ave}, and (d) Rku_{ave}.

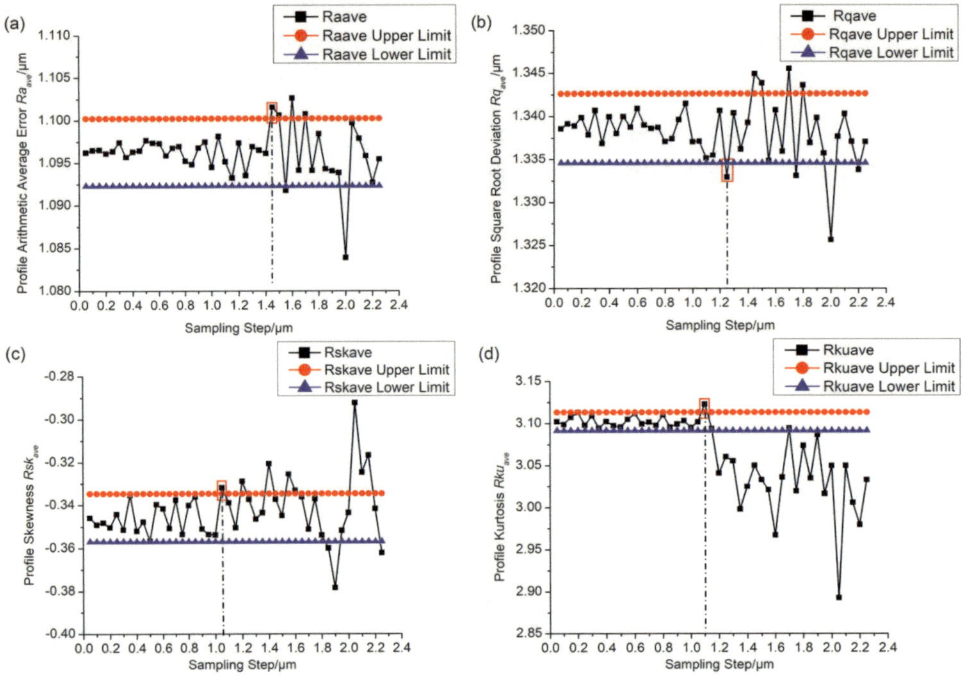

Figure 11.
Changing trends of four evaluation indexes with increasing sampling steps on fiber bundle end surface [27]. (a) Ra_{ave}, (b) Rq_{ave}, (c) Rsk_{ave}, and (d) Rku_{ave}.

Fiber bundle surface	Ra_{ave}/μm	Rq_{ave}/μm	Rsk_{ave}	Rku_{ave}
Side surface, ground	1.97 ± 0.03	2.44 ± 0.03	−0.55 ± 0.05	3.29 ± 0.10
Side surface, sandpaper polished	0.98 ± 0.02	1.41 ± 0.05	−2.15 ± 0.09	6.48 ± 0.59
Side surface, friction	0.56 ± 0.02	0.78 ± 0.02	−0.84 ± 0.07	7.96 ± 0.44
End surface, ground	1.23 ± 0.02	1.56 ± 0.03	−0.31 ± 0.04	3.60 ± 0.12
End surface, sandpaper polished	0.26 ± 0.01	0.36 ± 0.01	−1.39 ± 0.35	12.80 ± 0.55
End surface, friction	0.20 ± 0.01	0.24 ± 0.01	−0.34 ± 0.19	21.14 ± 2.88

Table 1.
Fiber bundle surface parameters of three processing methods [27].

After measuring the six surfaces with the sampling length of 280 μm and sampling number of 200, the scanning track perpendicular to the fiber orientation and machining direction, and the sampling step of 0.5 μm, four indexes can be calculated for every surface. The results are shown in **Table 1**.

The data in **Table 1** reflect that, regardless of the surface processing method used, the side surfaces are rougher than the end surfaces, which can be indicated by all side surfaces having larger Ra_{ave} and Rq_{ave} than the end surfaces. This phenomenon can be explained as follows. During the machining process, the anti-shear strength of the end surface is stronger; it is thus not easy for the fibers to be pulled-out and form surface damage, and the end surface becomes smoother than the side surface. However, fiber debonding and fiber delamination are more likely to appear, leading to more damage and a rougher surface on the side surface. In addition, for both fiber orientation surfaces, Ra_{ave} and Rq_{ave} of the friction-applied surfaces are the smallest, followed by sandpaper-polished surfaces, and finally ground surfaces. A conclusion can be made that Ra_{ave} and Rq_{ave} are both valid in evaluating the degree of surface roughness. The rougher the surface is, the larger Ra_{ave} and Rq_{ave} are.

As a comparison, Rku_{ave} of either a ground side surface or a ground end surface is roughly equal to 3. That is, the height distribution approximately obeys a Gaussian distribution on both surfaces. Polishing with sandpaper or sliding against a ZrO_2 disk can make the surfaces flat, thus decreasing the amount of surface damage. Meanwhile, they show larger Rku_{ave} values. It is clear that the surfaces after friction turn out to have the fewest numbers of surface defects, and Rku_{ave} of the friction surfaces has the biggest value among the three processing methods on both the side and end surfaces. It can be seen that Rku_{ave} is related to the amount of surface damage. The less damage a surface has, the larger Rku_{ave} is.

For both ground surfaces, Rsk_{ave} is close to 0, which is consistent with their Gaussian distribution characteristic. For the surfaces polished by a sandpaper, the crests are chipped off during this processing, with the original troughs remaining, and thus it is reasonable for these two surfaces to have a larger negative Rsk_{ave}. After friction is applied, however, their Rsk_{ave} values reach closer to 0 again, which can be explained by the wear debris embedded into the troughs during the friction

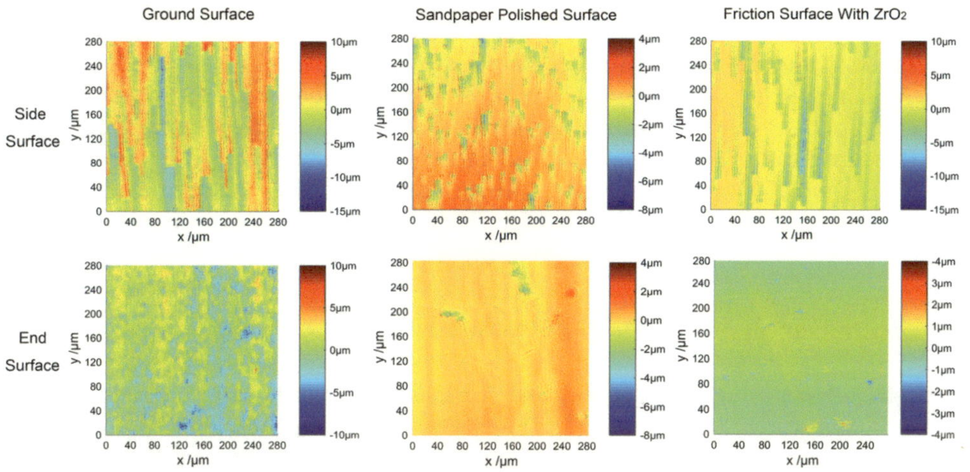

Figure 12.
Topographies of the six surfaces [27].

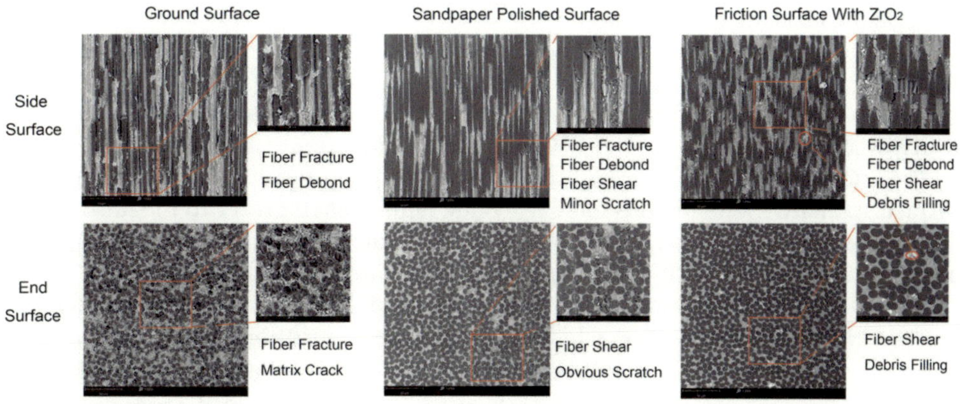

Figure 13.
SEM images of the six surfaces [27].

process decreasing the height of the troughs. Thus, it can be inferred that Rsk_{ave} is able to reflect the damage type or degree of the surface. A larger negative Rsk_{ave} value is caused by a trough-dominant surface, and a larger positive Rsk_{ave} value is caused by a crest-dominant surface. The surface state can be inferred by combining Rsk_{ave} and Rku_{ave}.

Figure 12 shows the fiber bundle surface topographies of three processing methods. **Figure 13** shows the microscopic surface topography of six surfaces. It is clear that the values of the four proposed indexes have a strong and direct connection with the surface damage and, thus, have a good feasibility and interpretability for a surface evaluation.

4. The measurement and evaluation of cell body surfaces and the whole surfaces

4.1 Measurement strategy

Cell body contains different fiber bundle orientations, and the whole surface is composed of cell bodies; thus standard procedures designated to 2D profile sampling at fiber bundle are in general not applicable for 3D topography measurement for cell body and the whole surfaces, because 2D measurement is of directionality which mainly reflects the damage between fibers, the fiber, and the matrix. Whereas for cell body and the whole surfaces, the scales are bigger, thus the measurement and evaluation mainly reflect the damage between fiber bundles and the matrix, which cannot consider the integrated effect of the fiber orientation and the processing direction simultaneously. Therefore, a 3D surface measurement and evaluation method should be adopted at these two grades.

4.2 Determination of MaxSS on cell body surfaces

A proper sampling step can make the surface information of a cell body extracted accurately and meanwhile save the cost of data collection and processing. MaxSS refers to the balance point of the accuracy and sampling cost. If a sampling step larger than the MaxSS is adopted, the information of a surface is distorted; if a sampling step larger than the MaxSS is adopted, unnecessary data sampling cost is spent. Therefore, how to determine the MaxSS on a cell body surface is very important. This section proposes a method for this topic, based on the principle of residual estimation.

For a cell body surface, the following steps can be executed to determine the MaxSS:

Sample the surface using a small sampling step of the measurement device which is at least one third of the WCMC fiber diameter.

Calculate *Sa*, *Sq*, *Ssk*, and *Sku* based on the sampling results, and set them as surface standard values θ (the standard value here does not refer to the ideal measurement result which has no error, rather, it means a standard which can be used to check whether other measurement results are acceptable).

Generally speaking, this standard value θ, containing measurement error, can be regarded as a random variable, which obeys Normal distribution, so, $\theta \sim N(\mu_1, {\sigma_1}^2)$. A measurement result $\overline{\theta}$ that is obtained from a larger sampling step obeys Normal distribution as well, so, $\overline{\theta} \sim N(\mu_2, {\sigma_2}^2)$. Because θ and $\overline{\theta}$ are both the measurement results of the same surface, their expectation is equal to the ideal real value (with no errors) of the surface. Therefore,

$$\mu_1 = \mu_2 \tag{15}$$

Set e as the difference between $\overline{\theta}$ and θ,

$$e = \overline{\theta} - \theta \tag{16}$$

Since $\overline{\theta}$ and θ are independent identically distributed (IID), their difference e obeys Normal distribution too:

$$e \sim N\left(\mu_2 - \mu_1, {\sigma_1}^2 + {\sigma_2}^2\right) \tag{17}$$

Combining Eqs. (16) and (17) together,

$$e \sim N\left(0, {\sigma_1}^2 + {\sigma_2}^2\right) \tag{18}$$

Based on the analysis above, for a set of measurement results obtained from different sampling steps, the Residual Errors (*RE*s) between each of them and the standard value θ are IID to Normal distribution, so

$$RE \sim N\left(0, \sigma^2\right) \tag{19}$$

For every actual engineering question, it is reasonable to find an acceptable range of *RE* according to the actual requirements of measurement. For example, if the $\pm 15\%$ smallest *RE*s are acceptable, the acceptable measurement results fall into the range of

$$[\theta\text{-}0.39\sigma, \theta + 0.39\sigma] \tag{20}$$

When the sampling step is small enough, the measurement result is in the range above. However, if the sampling step grows larger, the measurement result will go out of the range sooner or later. The largest sampling step that holds the measurement result within the range of Eq. (20) can be defined as the MaxSS for cell body surface measurement.

Here we take the measurement of a cell body of the C_f/SiC with processing angle of 90° as an example. The sampling steps of 1–45 μm were adopted to measure the cell body. The measurement results of 1 μm were set as the standard values. The rest of the results were compared with the standard values to calculate the *RE*s. The acceptable ranges of measurement results are available through Eq. (20).

The changing trends of the measurement results of the four evaluation indexes under different sampling steps are illustrated in **Figure 14**. The red and blue lines refer to the boundaries of the acceptable ranges calculated from Eq. (20). It is clear that when the sampling steps exceed a certain value (in red blocks), the corresponding results begin to go out of the ranges. Then the MaxSS can be determined for each index. Combining the four MaxSSs together, the MaxSS is available and for this material, it is 7 μm.

It can also be proved that the MaxSS of a cell body is approximately equal to the diameter of its reinforcing fiber.

4.3 Relationship between the measurement of cell body and the whole surface

The whole surface of a WCMC consists of many cell bodies. Some cell bodies nearby each other faced the same fabrication and machining process and may perform similar surface quality. Therefore, the measurement and evaluation of a

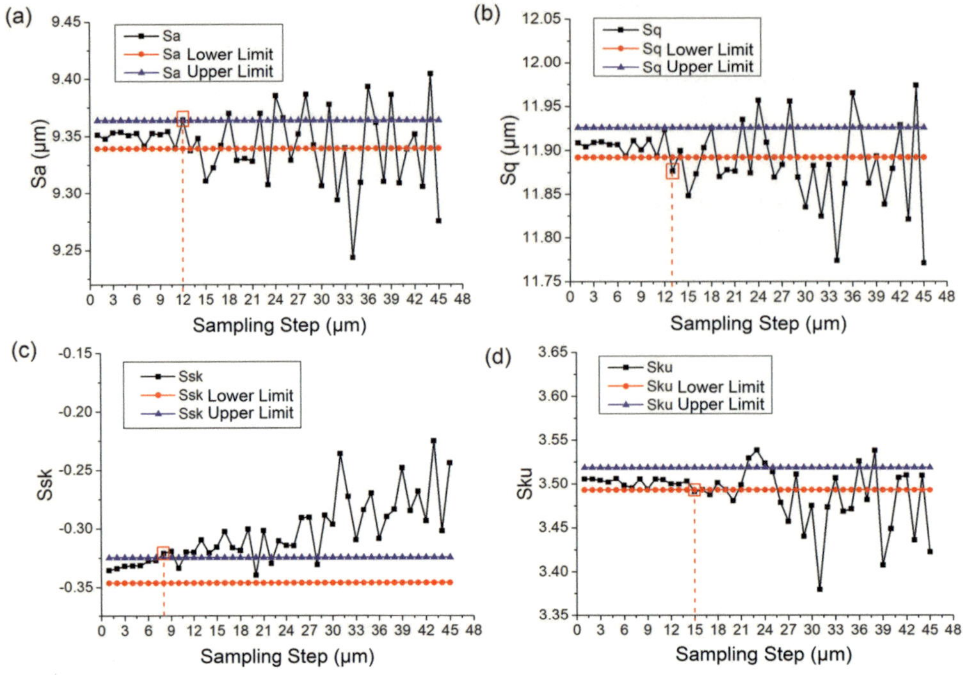

Figure 14.
The changing trends of (a) Sa (b) Sq (c) Ssk and (d) Sku with sampling steps on cell body surface [28].

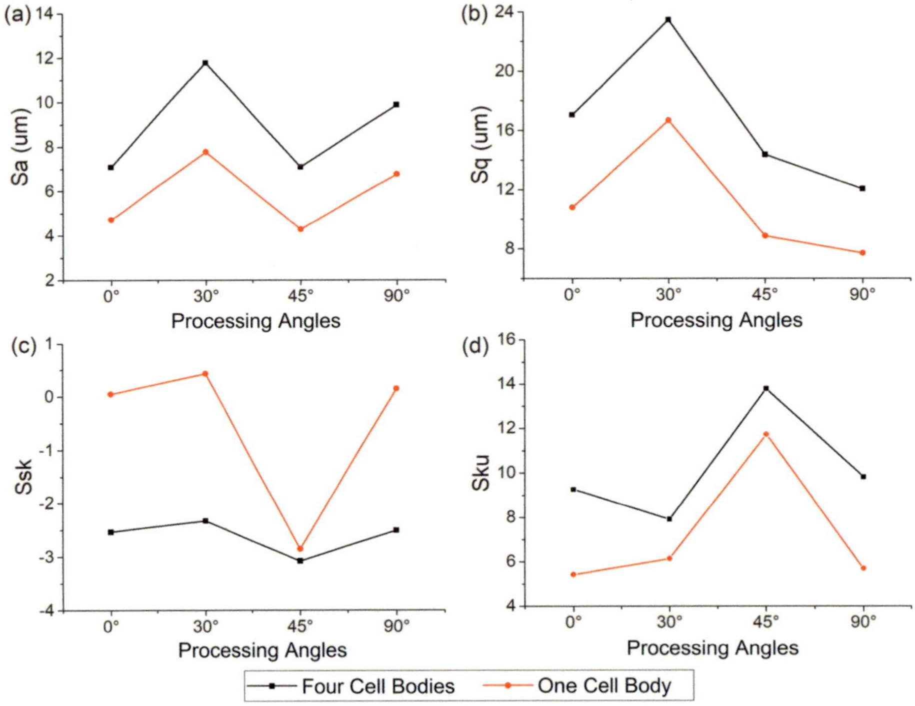

Figure 15.
Evaluation parameters with processing angles on cell body surface of C_f/SiC (a) Sa, (b) Sq, (c) Ssk, and (d) Sku.

cell body can be used to estimate the surface quality state of a certain area nearby it. It has been proved that, for the exampled C_f/SiC with processing angles of 0°, 30°, 45°, and 90°, the measurement results of the four indexes of one cell body have the similar values with the results of the nearby four cell bodies (shown in **Figure 15**).

5. Conclusions

This chapter aims at providing a grading surface measurement and evaluation system for woven ceramic matrix composites. The system contains four grading of fiber, fiber bundle, cell body, and the whole surface. The main conclusions are as follows:

1. The type and degree of the damage on fibers influence the processing quality and property of the surface. The diameter and the direction of the fibers determine the measurement parameters when sampling fiber bundle or cell body surfaces.

2. 2D measurement should be adopted on fiber bundle surfaces. Sampling parameters, including sampling length, number, step, and direction should be determined carefully to balance the accuracy and the efficiency. Four evaluation indexes, namely, *Ra*, *Rq*, *Rsk*, and *Rku*, are usable for fiber bundle surface evaluation.

3. 3D measurement should be adopted on cell body surfaces. Maximum sampling step can be determined with the principle of residual estimate. *Sa*, *Sq*, *Ssk*, and *Sku* are usable on this grade.

4. The whole surface is consist of many cell bodies. Therefore, a small number of cell bodies can be used to represent a larger area nearby. This idea can help reduce the workload when measuring and evaluating a large area of WCMC surface.

Acknowledgements

Special thanks the National Natural Science Foundation of China (Nos. 51375333 and 51805366) for financial assistance.

Conflict of interest

All the authors listed have approved the manuscript, and no interest of any third parties is infringed.

Author details

Bin Lin*, Haoji Wang and Jinhua Wei
Key Laboratory of Advanced Ceramics and Machining Technology of Ministry of Education, Tianjin, China

*Address all correspondence to: linbin@tju.edu.cn

References

[1] Söderfjäll M, Herbst HM, Larsson R, Almqvist A. Influence on friction from piston ring design, cylinder liner roughness and lubricant properties. Tribology International. 2017;**116**: 272-284

[2] Erol O, Powers BM, Keefe M. Effects of weave architecture and mesoscale material properties on the macroscale mechanical response of advanced woven fabrics. Composites Part A: Applied Science and Manufacturing. 2017;**101**:554-566

[3] Hosseini Monazzah A, Pouraliakbar H, Bagheri R, Seyed Reihani SM. Al-Mg-Si/SiC laminated composites: Fabrication, architectural characteristics, toughness, damage tolerance, fracture mechanisms. Composites Part B: Engineering. 2017; **125**:49-70

[4] Cao HM, Zhou X, Li XY, Lu K. Friction mechanism in the running-in stage of copper: From plastic deformation to delamination and oxidation. Tribology International. 2017;**115**:3-7

[5] Krenkel W, Heidenreich B, Renz R. C/C-SiC composites for advanced friction systems. Advanced Engineering Materials. 2002;**4**:427-436

[6] Manocha LM, Prasad G, Manocha S. Carbon-ceramic composites for friction applications. Mechanics of Advanced Materials and Structures. 2014;**21**: 172-180

[7] Sullivan P. J and blunt L, Three-dimensional characterization of indentation topography: Visual characterization. Wear. 1992;**159**: 207-221

[8] Senin N, Ziliotti M, Groppetti R. Three-dimensional surface topography segmentation through clustering. Wear. 2007;**262**:395-410

[9] Zhao F. 3D evaluation method of cutting surface topography of carbon/ phenolic (C/Ph) composite. Journal of Wuhan University of Technology—Materials Science Edition. 2011;**26**: 459-463

[10] Hintze W, Cordes M, Koerkel G. Influence of weave structure on delamination when milling CFRP. Journal of Materials Processing Technology. 2015;**216**:199-205

[11] Cao X, Lin B, Wang Y, Wang S. Influence of diamond wheel grinding process on surface micro-topography and properties of SiO_2/SiO_2 composite. Applied Surface Science. 2014;**292**: 181-189

[12] Cao X, Lin B, Zhang X. Investigations on grinding process of woven ceramic matrix composite based on reinforced fiber orientations. Composites Part B: Engineering. 2015; **71**:184-192

[13] Zhao F-L, Al C-Z, Yang D-J, Yang Z-X, Wang J-M, Ao M. Study on the evaluation method and evaluation parameters of cutting surface roughness of carbon/carbon composite. Acta Metrologica Sinica. 2006;**27**:206-211

[14] Senin N, Ziliotti M, Groppetti R. Three-dimensional surface topography segmentation through clustering. Wear. 2007;**262**:395-410

[15] Hocheng H, Tai N. H and Liu C S, Assessment of ultrasonic drilling of C/ SiC composite material. Composites Part A: Applied Science and Manufacturing. 2000;**31**:133-142

[16] Tashiro T, Fujiwara J, Takenaka Y. Grinding of C/C-SiC Composite in Dry Method. London: Springer; 2007

[17] Xu W, Zhang LC. On the mechanics and material removal mechanisms of

vibration-assisted cutting of unidirectional fibre-reinforced polymer composites. International Journal of Machine Tools and Manufacture. 2014; **80–81**:1-10

[18] Zhang L, Ren C, Ji C, Wang Z, Chen G. Effect of fiber orientations on surface grinding process of unidirectional C/SiC composites. Applied Surface Science. 2016;**366**: 424-431

[19] Fabre D, Bonnet C, Rech J, Mabrouki T. Optimization of surface roughness in broaching CIRP. Journal of Manufacturing Science and Technology. 2017;**18**:115-127

[20] Bian R, He N, Ding W, Liu S. A study on the tool wear of PCD micro end mills in ductile milling of ZrO_2 ceramics. The International Journal of Advanced Manufacturing Technology. 2017;**92**:2197-2206

[21] Cao X. Lin B and Zhang X, A study on grinding surface waviness of woven ceramic matrix composites. Applied Surface Science. 2013;**270**:503-512

[22] Chou T-W, Ko FK. Textile structural composites. In: Composite Materials Series. Vol. 3. Amsterdam, New York, USA: Elsevier Science Publishers; 1989

[23] Mouritz A, Baini C, Herszberg I. Mode I interlaminar fracture toughness properties of advanced textile fibreglass composites. Composites Part A: Applied Science and Manufacturing. 1999;**30**: 859-870

[24] Wei J, Lin B, Cao X, Zhang X, Fang S. Two-dimensional evaluation of 3D needled Cf/SiC composite fiber bundle surface. Applied Surface Science. 2015;**355**:166-170

[25] Fan SW, Xu YD, Zhang LT, Cheng LF, Yu L, Yuan YD, et al. Three-dimensional needled carbon/silicon carbide composites with high friction performance. Materials Science and Engineering: A. 2007;**467**:53-58

[26] Wei J, Lin B, Wang H, Sui T, Yan S, Zhao F, et al. Friction and wear characteristics of carbon fiber reinforced silicon carbide ceramic matrix (Cf/SiC) composite and zirconia (ZrO_2) ceramic under dry condition. Tribology International. 2018;**119**:45-54

[27] Wei J, Wang H, Lin B, Sui T, Wang A, Zhao F, et al. Measurement and evaluation of fiber bundle surface of long fiber reinforced woven composites. Surface Topography: Metrology and Properties. 2019;**7**:015003

[28] Wei J, Wang H, Lin B. Measurement of cell body and the whole surfaces of long fiber reinforced woven composites. In: IOP Conference Series: Materials Science and Engineering. Vol. 678. 2019. p. 012029